Design for Manufacturing and Assembly

Concepts, architectures and implementation

O. Molloy
Nortel
Galway
Republic of Ireland

S. Tilley
WTCM-CRIF Research
Heverlee
Belgium

and

E. Warman
K. Four Ltd
Peterborough
UK

CHAPMAN & HALL
London · Weinheim · New York · Tokyo · Melbourne · Madras

Published by Chapman & Hall, an imprint of Thomson Science, 2–6 Boundary Row, London SE1 8HN, UK

Thomson Science, 2–6 Boundary Row, London SE1 8HN, UK

Thomson Science, 115 Fifth Avenue, New York, NY 10003, USA

Thomson Science, Suite 750, 400 Market Street, Philadelphia, PA 19106, USA

Thomson Science, Pappelallee 3, 69469 Weinheim, Germany

First edition 1998

© 1998 Chapman & Hall

Thomson Science is a division of International Thomson Publishing

Printed in Great Britain by T J International Ltd, Padstow, Cornwall

ISBN 0 412 78190 5

A catalogue record for this book is available from the British Library

Owen Molloy dedicates this book to his wife Rosaleen, for her patience and support.

~

Steven Tilley would like to dedicate the book to his parents for giving him the opportunity to work in the fascinating world of technology.

~

Ernie Warman wishes to dedicate this book to his family for their patience and to the memory of Jozeph Hatvaney for inspiring him.

Contents

4. The Product Model and CAD Interfacing   **98**

5. Knowledge Engineering and Inferencing   **117**

About the Authors

Owen Molloy

Owen Molloy is a Senior Software Engineer with Nortel in Galway, Ireland. Previously he worked as project manager in the area of Concurrent Engineering at the CIM Research Unit, Galway, Ireland, where he managed a number of Irish and European industrial research projects in Concurrent Engineering and Design for Manufacturing and Assembly. These included the BRITE EuRam DEFMAT (Design for Manufacturing Architecture and Tool Suite) project, which led to the production of this book. Previously he worked as a software engineer for Dassault Systemes, France, working in the area of CAD Interfacing. He obtained his PhD in Industrial Engineering from University College, Galway in 1995. He holds an MEngSc degree in Electronic Engineering from University College, Dublin.

Steven Tilley

Steven Tilley (1964) finished his degree in Mechanical Engineering at the Katholieke Universiteit Leuven (Belgium) in 1988 on the off-line programming of robots. Since 1988 he has been a research engineer with WTCM, the research centre of the metal working industry in Belgium. In the period 1988-1990 he was involved in several national and European projects in the area of Process Planning for sheet-metal products : PRESS, BRITE 2406. In the period 1990-1995 he was involved in Design For X projects: Design to Cost for sheet-metal, IDAM (Integrated Design for Assembly and Manufacturing), BRITE DEFMAT (Design for Manufacturing and Assembly Architecture) (4661) and Esprit CONSENS (CONcurrent Simultaneous ENgineering System) (6896). Since 1995 he has been involved in the development of new design and process planning methods for sheet-metal products and production methods, specifically the design and implementation of software tools for the selection of tools on punching machines and the sequencing and quality assurance on bending machines.

Ernie Warman

Ernie Warman initially trained in the Automotive Industry at F.Perkins Ltd., after which he graduated in Production Engineering and Mechanical Engineering. Following completion of his Ph.D, he was promoted to be totally responsible for the technical computing and automation strategy of the Perkins Engines Group (10,000 employees) and relating this strategy to the overall requirements of Massey Ferguson (the parent company). He co-founded Productivity International in Dallas Texas, where work was undertaken for major corporations in introducing new design and manufacturing technologies and the required organisational changes. Following this he joined Inbucon Consultants as a Senior Associate where he established their presence in computer aided design and integrated manufacturing. He then founded K.Four Ltd. and a wide range of work relating to design, manufacturing and organisational impact has been undertaken for leading companies. He acts as a project reviewer for many EEC projects, Euromanagement consultant for DG XXIII of the EC, and reviewer for ACME directorate of the SERC on design and manufacturing research projects. Dr Warman has published over 60 papers and is on the editorial board of Computers and Industry and has extensive professional activities with the various engineering bodies of which he is a member. He is a member of I.Mech.E., I.Manf.E., I.E.E., B.C.S., and A.C.M. He is a recipient of the IFIP Silver Core award.

Preface

In today's economic environment, companies must produce greater product variety, at lower cost, all within a reduced product life-cycle, in order to compete and survive. In order to achieve these goals, a Concurrent Engineering approach to the design process must be adopted, where concurrent consideration of life-cycle constraints leads to a 'right first time' approach to product design. Central to the Concurrent Engineering philosophy, and in many cases the main realisation of it, is Design For Manufacture and Assembly (DFM/A). It is to DFM/A that companies most often look for immediate benefits of an integrated approach to product and process design.

To reap the full rewards of DFM/A, companies should be able to assess their products with respect to their own manufacturing processes and equipment. Further considerations include the ability to analyse product designs against various process and equipment combinations, to assess cost and manufacturing lead time considerations. Overall a DFM/A system needs to answer manufacturing cost as well as design functionality questions and give these constraints their proper "importance" weightings. A move to a state of "reasoning with knowledge" can be argued to be the best approach for these DFM/A traits. The knowledge-based approach enables, for example, a repeated cycle of process selection, cost estimation and design evaluation. This move is all the more justified since the advent of computer technology which offers fast, efficient processing of knowledge and data as well as a platform for a direct link between knowledge-based systems and CAD systems.

Effective DFM/A must take place through feedback to the design domain. Introducing multiple DFM/A criteria into the design process raises numerous issues, such as how to incorporate DFM/A considerations into design without seriously retarding the design process. It will also be more difficult to optimise a design with respect to many (often conflicting) criteria, and to weigh these criteria against each other. If DFM/A is to be automated it must be possible to relate particular DFM/A criteria to particular design entities, even from the conceptual design stage. It should also be possible to filter the DFM/A analysis to coincide with a company's design review process, so that the appropriate analyses are carried out at each review stage. This would ensure completeness as well as accountability.

There is a need for new architectures for DFM/A systems which capitalise on the latest software and knowledge based techniques to deliver the DFM/A systems of tomorrow. Such architectures must be based upon complete understanding of the issues involved in integrating the design and manufacturing domains.

This book provides a comprehensive view of the capabilities of advanced DFM/A systems based on an advanced generic DFM/A systems architecture. It addresses the potential for improvement of currently available DFM/A solutions and points the way forward through adoption of a comprehensive architecture based approach to the design of DFM/A systems.

Acknowledgements

The authors wish to thank a number of people, without whose support this book would not have been possible. The initial DEFMAT concept was formulated by Jos Pinte and Prof. Jimmie Browne. We wish to give our special thanks to the European Union DG XII and specifically to the DEFMAT project officer Antonio Colaço. A large number of people were involved in the DEFMAT project during it's three and a half years intensive research and development work, contributing their ideas and hard work to make the concept a reality. Notable among these were Ip Shing Fan, Steffen Neu, Stephan Kruger, Martin Devilly, Harold Rothenberg, Cathal Gallagher, Lorcan Mannion, George Brophy and numerous others from Digital, IWF, WTCM, CIMI, CIMRU and AEG who we hope will forgive not being named individually. In particular we wish to thank Jarek and Grazyna Martusewicz, Pieter Kesteloot and Patrick Meylemans for their participation and support during the design and implementation of the CAD interface and the integration phase of the software.

The authors gratefully acknowledge the support of the DEFMAT research work by the Commission of the European Communities, under the BRITE/EURAM program, contract number BREU-CT 91-0492, proposal number 4661.

1. Design For Manufacture and Assembly Concepts

There are a number of major pressures currently making themselves felt in manufacturing industry, coming both from customers and the business environment.

Globalisation of the marketplace (Browne, 1992), due to such factors as accessibility of markets and improvements in transport, is forcing manufacturers to operate in the context of global standards. Great emphasis is placed on becoming *world class*. In order to become more competitive a common strategic priority is to develop integrated engineering and manufacturing systems that address shorter product development cycles, increased product quality and reduced product cost. There is a shift from mass production towards make-to-order and ultimately one-of-a-kind production, with emphasis being placed on product range, variation, customisation, quality and cost.

There is a growing tendency towards the "Extended Enterprise" (Browne, 1992), where different enterprises supply expertise and capability in different areas (e.g. design, manufacturing, distribution, marketing) and co-operate to exploit a business (product) opportunity. This creates a need for new organisational structures and tools to support multi-functional design incorporating a life-cycle view throughout the design process.

Environmentally benign production has become a key issue, from the legislative, business and customer points of view. Recyclability and environmental costing are now important design issues. According to Tipnis (1993, 1994), a new paradigm *"Green Products plus lean Production equals Sustainable Growth"* has emerged to which responsible companies must respond to compete and survive. Issues such as Design for Recycling and the environment must be studied in the context of the changing value-chain and the extended enterprise, and its effect on competitivity.

The trend towards reduced manufacturing product life, coupled with extended field life has been noted by Tomiyama (1992). Tomiyama argues that current mass-production methods are not sustainable on a global scale, and proposes a post mass-production era based on modular products with long life-cycles, which may be upgraded as well as individual components recycled. Such trends (which are already visible in the personal computer market) place greater pressure on manufacturers to plan for product life-cycle phases such as disassembly, modularity, compatibility between product versions, and indeed disposal.

With the moves towards recycling in recent years (witness the German automotive industry's push towards totally recyclable cars), Design For Disassembly is becoming a major influence on certain products (Boothroyd, 1992). Weule (1993) reports on the Daimler-Benz group's method of life-cycle analysis, whose principle is to measure the impact on the ecosystem of all aspects of the product life-cycle. The increases in the rate of scrapped products, stemming partly from better distribution, cheaper products, and shorter lifetimes, have also meant the tightening of regulations regarding the recyclability of products. Emerging policies reported by Jovane (1993) include:

1. incentives for manufacturers who use recycled materials when these are not the most economic choice
2. taxes on the use of virgin materials, as well as deposit fees to be returned when these materials are to be recycled
3. systems to measure recyclability of products, so that manufacturers could be penalised if they do not meet specific recyclability standards.

These multiple criteria of life-cycle analysis (e.g. manufacture, test, maintenance, environment, recycling) may be grouped under the term Design For X (DFX). DFX incorporates the manufacturing and assembly criteria for design, as well as beyond to the rest of the product life-cycle.

Such pressures as noted above have raised the need to adopt a life-cycle view of products through all design stages. This life-cycle view is commonly called Concurrent Engineering (CE) or Simultaneous Engineering (SE). Of course CE means different things to different people - strategy, philosophy, methodologies. For example, Canty (1987), provides a very broad strategy statement from Digital

Equipment Corporation: "CE is both a philosophy and an environment. As a philosophy, CE is based on each individual's recognition of his/her own responsibility for the quality of the product. As an environment it is based on the parallel design of the product and the processes that affect throughout its life-cycle" (Figure 1.1). Through addressing life-cycle issues earlier in the design process, it is intended to gain cost and time-to-market benefits (Figure 1.2). Other commentators such as the UK CE industrial forum, and Eversheim and Gross (1990), give more focused definitions of CE, which accentuate the need to produce better products faster and cheaper, using parallel task processing to shorten time to market.

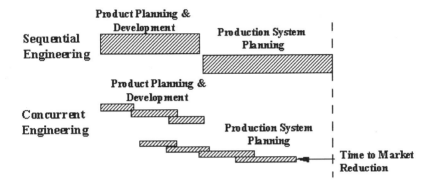

Figure 1.1 *Early Life-Cycle Planning Through Concurrent Engineering*

What is clear is that design of products and their associated processes must happen concurrently in order to reap the greatest benefits from taking the life-cycle view of the product.

In itself, the term "Concurrent Engineering" places no limits on the number of possible interpretations, and is really an aspiration: to appraise and design all aspects of a product and its life-cycle concurrently in order to reach a solution satisfying all these aspects and their inter-relationships; a very broad brief, if it is to be fully embraced.

The USA Defence Advanced Research Projects Agency (DARPA) Initiative in CE (DICE) states (DARPA, 1990): "CE is a systematic approach to the integrated, concurrent design of products and their related processes, including their manufacture and support. This approach is intended to cause the developers, from the outset, to consider all elements of the product life-cycle from conception through disposal, including quality, cost, schedule and user requirements".

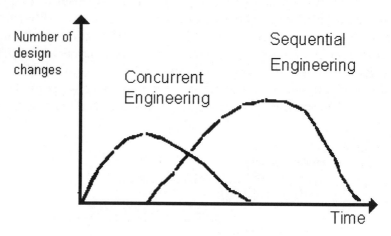

Figure 1.2 *Early CE Means Fewer Late Design Changes*

DARPA also concluded (Reddy *et al.*, 1992) that "..advanced computer software to assist a human team in considering all aspects of a product, including manufacture and logistical support, concurrently from the outset, is essential for the development of high-quality products in the shortest possible time at affordable costs". The mission of the DICE initiative (Engines, 1989) is to create an open computer-assisted CE environment, consisting of:

1. a shared information model that captures complete descriptions of the product or system and all associated process activities and organisational activities
2. a global object framework that enables the use of the shared information model by a network of co-operating computer-based clients
3. services, methods, tools and advisors that assist concept evaluation, analysis and intelligent decision making.

The stated aim of DICE is to emulate, for large organisations, the tiger-team approach to CE, which is successful for small groups. Indeed CE is fundamentally about taking a team approach to product development, and the necessary organisational and cultural adaptations must be understood before supporting technology can be introduced. The problem of migration from a sequential development environment to the implementation of CE architectures is the subject of considerable research (e.g. Singh (1992), Bradley and Molloy (1995)). The adaptation of Business Process Redesign (BPR) tools to develop CE strategies may in future help in this migration.

In 1990 Japan initiated the IMS (Intelligent Manufacturing Systems) international collaborative project with the US, Europe, Australia and Canada. Prompted by population trends leading to fewer people entering manufacturing, Japan identified the need for R&D in the areas of integration, production technology and systems. The unity of design and manufacturing is explicitly recognised in the IMS proposal.

1.1 Implementation of Concurrent Engineering

The implementation of CE may be approached in a number of complementary ways, such as:

1. integration and optimisation of product and process design
2. organisational changes to promote parallel product and production planning
3. the adoption of a team approach to product development
4. technological approaches to improvement of communications and data sharing facilities.

In practice, companies and individuals tend to tackle CE from particular perspectives, employing a variety of methodologies used either by multidisciplinary product development teams or by individuals. These techniques range from generic methodologies such as Design For Manufacture / Assembly (DMF/A) guidelines, Failure Modes and Effects Analysis (FMEA) and Quality Function Deployment (QFD) to specific company guidelines and tools. Because Concurrent Engineering implies interaction between functions, personnel and departments which hitherto tended to be more isolated, improved communication procedures and computer-based tools must be provided to enable CE to take place. Implementation of CE may necessitate new organisational structures, special training, investment in IT tools to support team design and information management for CE.

The approach to DFM/A proposed in this book is to provide fast, company-specific support to the resolution of the conflicting requirements imposed by manufacturing and assembly during the different stages of design, while minimising the dilution of design requirements imposed by such additional analysis methods. However the inherent advantage of addressing DFX constraints earlier rather than later in the design process cannot be understated. Wallace and Suh (1993) have developed a computer program to derive design strategies for products based on customer requirements. The basic principles used here could be taken a step further by integrating a number of analysis tools (such as QFD, FMEA, DFM/A) through a common product model, as proposed by Molloy (1995). His prototype system (Figure 1.3) demonstrated the use of an expert systems-based approach to a CE design system allowing the concurrent resolution of both customer and manufacturing constraints. As noted by Ishii *et al.* (1993a), DFX goals are going to conflict, requiring human intervention to weigh out the various options.

Effective DFX must take place through feedback to the design domain. Introducing multiple DFX criteria into the design process raises numerous issues, such as how to incorporate DFX considerations into design without seriously retarding the design process. It will also be more difficult to optimise a design with respect to many (often conflicting) criteria, and to weigh these criteria against each other. If DFX is to be automated it must be possible to relate particular DFX criteria to particular design entities, even from the conceptual design stage. It should also be possible to filter the DFX analysis to coincide with a company's design review process, so that the appropriate analyses are carried out at each review stage, perhaps using hypermedia solutions to present relevant information to designers (Spath, 1994). This would ensure completeness as well as accountability.

DFX analysis of a product may take place on several levels. Alting and Jorgensen (1993) point out that environmental screening of a product may take place on the four levels of:

1) product function,
2) product structure,
3) product life-cycle and
4) product components,

and proposes a procedure for this screening process on each of these levels. For example, identification of the essential parameters for product function will lead to a comparison of the actual design, and theoretical minimum consumption of materials and energy to achieve the desired functionality.

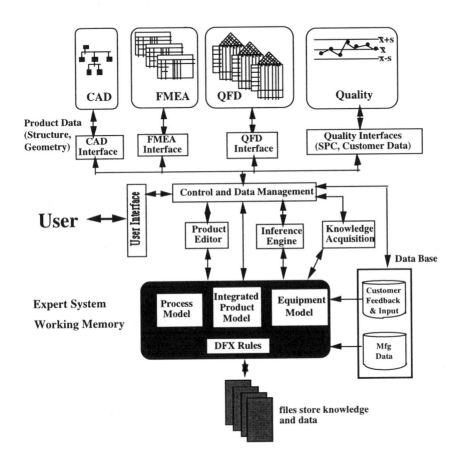

Figure 1.3 *CE Design Environment Proposal (Molloy, 1995)*

Essential to a useful quantitative evaluation using Alting's procedure is the ability to directly relate particular product characteristics to environmental loads. Such relationships can be established as constraints on the design, and used during the design process. Such analysis could be performed using an adapted QFD procedure. However a more quantitative evaluation is also desirable, especially as environmental effects must often be measured against accepted levels. Similarly, product DFM/A may be applied at the level of product range (Browne *et al.*, 1985). The manufacturing process chosen may have a direct bearing on the product functionality and it is of prime importance in the analysis of the product (e.g. when deciding when integration of parts is possible).

1.2 Issues involved in introducing DFM/A

Given that there is a need for DFM/A, the following questions may help clarify the issues involved before an implementation is undertaken in a company:

1. What are the available DFM/A knowledge sources (both internal and external)? How can the knowledge be acquired, organised and maintained in a form accessible by both design and manufacturing personnel, potentially in some form of DFM/A expert system?
2. Which product and process information are needed for DFM/A? How should this information be gathered, organised and accessed?
3. Is the final solution capable of being used at all stages in the product design, or is a finished design required? Obviously, the earlier in the design process the DFM/A solution can be applied, the more effective it will be.
4. What training will be required to integrate the DFM/A into the design process, and what will be the cost in terms of additional time and effort in using it, compared to the potential payback?
5. Is balance maintained between DFM/A, and Manufacture/Assembly for Design? For example it may be worthwhile for a company to look outside it's current manufacturing capability if a design requires a new process.
6. Is the DFM/A approach adopted seen by the designer as being over-constraining or by the manufacturing engineer as giving too much power to the designer in choosing the manufacturing parameters?

1.3 DFM/A Principles and Techniques

One of the primary techniques in DFX is Design For Manufacture (DFM), a concept which concentrates on the integration of manufacturing criteria into the product design process. The essence of the DFM approach is the integration of product design and process planning into one common activity (Stoll, 1988). DFM may provide the link between design and process planning. The most widely applied strategy in DFM to date is that of Design For Assembly (DFA), whose primary aim is to ensure that products are easier (hence cheaper) to assemble. DFA is often regarded as the key to successful product DFM (Boothroyd, 1993). The consequent reduction in part costs, after DFA evaluation, often outweighs the assembly cost reductions. Thus DFA is often seen as central to the issue of product DFM. Stoll (1988) presents the most common DFM/A methodologies. Among the supporting techniques is Group Technology, which seeks to reduce manufacturing system information content by using codes to identify and exploit the similarity of parts. Group Technology can thus be regarded as an aid in the simplification of the DFM/A task as it does not contain any DFM/A analysis function), and the DFA Handbook, developed by Boothroyd and Dewhurst (1988). This latter was widely regarded as a major development in the formalisation of DFA thinking.

The designer must have information on the manufacturing methods and life-cycle of the product he is designing. Taking manufacturing, this is not a simple matter, as several different manufacturing processes may be used for a single product. The designer can only take into consideration the strengths and weaknesses of the manufacturing methods available and fully exploit them, if he has a proper source of manufacturing information, presented in a format which he can relate to his design. Three main classifications of DFM/A knowledge can be identified:

1. general guidelines
2. company-specific best practice
3. process and or resource-specific constraints

General guidelines refer to generally applicable rules-of-thumb, relating to a manufacturing domain, of which the designer should be aware.

These are usually expressed as a checklist for the designer. Some examples of general DFA guidelines are:

1. parts must be simplified and standardised as much as possible
2. parts should be symmetrical
3. fasteners should be minimised

Company-specific best practice refers to those in-house design rules which a company develops, usually over a long period of time, which the designer is expected to adhere to. These design rules are identified by the company as contributing to improved quality and efficiency by recognising the overall relationships between particular processes and design decisions. Such guidelines are used by companies as part of the training given to designers of products requiring significant amounts of manual assembly or maintenance.

Resource-specific constraints refer to situations where particular design parameters can be judged to be within acceptable range based on limitations imposed by particular processes. For example, in PCB assembly, the following rule may apply:

the minimum distance of components from the board edge must be greater than the conveyor belt width.

The value used for the conveyor width should be that of the actual machines used in manufacturing the product. For example, not all PCBs may be wave-soldered, and there may be different wave-solder machines used in different plants manufacturing the same product. Therefore rules and constraints must be related to the current process and equipment models relevant to the product being analysed.

The generic architecture proposed in this book allows for this through customisation of the DFM/A analysis. An example is given of the possibility of analysing a PCB assembly against two different assembly lines, with the same processes but slightly different machine parameters (see Section 6.11).

1.4   Current state of commercial DFM/A packages

Most of the currently available DFM/A techniques are spreadsheets, requiring the designer to answer questions relating to the product and its components, their form and functionality, and how they interact. Examples of these are the Lucas DFA technique (Corbett *et al.*, 1991), the Hitachi Assemblability Evaluation Method (Miyakawa and Ohashi, 1986), and the Boothroyd-Dewhurst methodology.

The general sequence of DFA analysis in the Lucas and the Boothroyd-Dewhurst methodologies is similar. The first step is the determination of the product and sub-assemblies assembly sequence. This provides the product structure as a basis for the following DFA analysis. Through a standard question and answer procedure, it is determined whether particular components are essential to the product, or can be avoided through, for example, integration with other parts or the use of different assembly techniques. Through definition of the shape, size and symmetry of each part and sub-assembly, the systems identify feeding, handling and insertion problems. They also allow the estimation of assembly times and costs.

The Boothroyd-Dewhurst DFA method is based on industrial time study methods. It aims to minimise assembly costs by reducing the number of parts to be assembled and then ensuring that the remaining parts are easy to assemble. Non-essential parts and time penalties for difficult operations reduce the overall "score" for design efficiency. Boothroyd and Dewhurst also supply a number of other DFM/A modules including sheet-metal and injection moulding analysis, as well as a module for interfacing with the Pro/Engineer CAD system to provide automatic interface with the CAD product structure.

In the Hitachi Assemblability Evaluation Method (AEM) for DFA analysis, the two criteria of ease of assembly (as a measure of design quality), and the estimated assembly cost are used to distinguish between designs. The AEM does not distinguish between the method of assembly used during the analysis, where a range of symbols are used to represent assembly operations. Hitachi claimed a deviation of ± 10 % in the estimated assembly cost from actual costs for small mass-produced products, with worsening accuracy as product size increases. Thus this method is very much aimed at a mass-production environment.

The current developments in DFM/A tools, and the contribution of complementary methodologies such as QFD and FMEA will be described in more detail in Chapter 2.

1.5   Requirements for a new generation of DFM/A systems

While the currently available DFM/A methodologies are widely used by industry and provide higher benefits when used in a CE team environment, it is possible to outline where improvements can be offered by more advanced IT-based analysis systems. Some of the areas for potential improvements are:

1. direct analysis of designs, through improved CAD interfaces, thereby not relying on all designers to answer all questions all the time in a consistent manner.

2. customisation of the analysis done on the product design to reflect the actual manufacturing concerns of the user, and to produce different results for different configurations of processes or equipment

3. more specific analysis of product designs, through relation of design features to manufacturing features and processes.

4. there is no mechanism for incremental capture of manufacturing rules and decisions.

1.6   Knowledge-based approaches to DFM/A

DFM/A systems which could deliver these functionalities would achieve a higher level of integration into the design and manufacturing functions of the company. In attempts to provide viable commercial solutions, research in DFM/A today is focusing on a knowledge-based approach to the design of DFM/A expert systems. A more detailed explanation of knowledge engineering for DFM/A is provided in Chapter 4. An early example is the prototype AI based design system CADEMA (O'Grady *et al.*, 1988). This system analyses design from a manufacturing and assembly viewpoint and recommends changes to designers to improve the functional and manufacturing aspects of the design (Chang, 1990). This and further research by the same team (Oh *et al.*, 1991), (Bowen and Bahler, 1991) is based on the use of a constraint networks to incorporate the DFX rules. The advantages of such languages are that they support multi-directional inference (Duffy *et al.*, 1994), and that one constraint may more easily encapsulate knowledge which would require several rules to express. They also provide easy ways of associating data and constraints within one representation format, with changes propagating throughout the network. However the ability of expert systems to separate knowledge and data is useful when domain experts wish to update models of equipment or DFM/A knowledge bases, without reprogramming the constraint network. With expert systems, the functionality of the system is not so tightly entertwined with the knowledge being applied.

The problem of DFM/A is amenable to solution using expert systems: it necessitates the systematic application of large amounts of knowledge, which is normally only acquired over long periods of time. The solutions to the problems of knowledge elicitation and acquisition have yet to be formalised into a generic approach and are usually separated from the knowledge modelling problem. According to Chang (1990), an expert system is a tool which has the capability to understand problem-specific knowledge and use the domain knowledge intelligently to suggest alternative paths of action. A typical expert system (Figure 1.4) consists of (1) a knowledge-base of domain-specific information, (2) some form of rule structure for inferencing and (3) an inference mechanism which acts upon the rule base.

The use of expert systems in DFM system design to provide the central inferencing mechanism is now commonplace, though as demonstrated in (Molloy and Browne, 1993) and (Brophy *et al.*, 1993), it is feasible to develop DFM systems for specific domains, for example PCB assembly, without recourse to expert systems software. Both the DEFMAT system implementations (see Chapter 6) and the DFM system described by Venkatachalam *et al.* (1993) use the Nexpert™ expert system shell as the main inference engine. Rules-based AI systems provide powerful rule and object-based knowledge and data modelling possibilities (Figure 1.5).

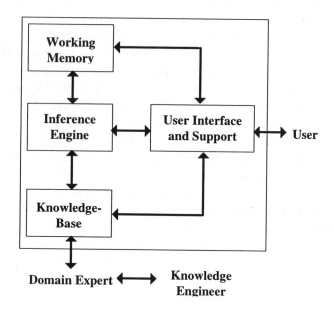

Figure 1.4 *Generic Expert System Model*

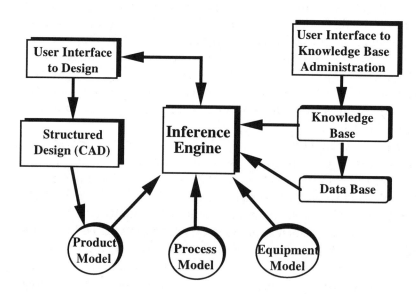

Figure 1.5 *Expert Systems Approach to Design for Manufacture and Assembly*

1.7   Interfacing Design (CAD) and DFM/A Systems

The goal of many DFM/A systems is to provide direct interfaces with CAD tools. Modern CAD systems vary from 2D draughting type systems to full 3D capabilities, up to parametric feature-based design systems. The differences in the geometric modelling capabilities of the different systems are reflected in their applications: 2D systems are well suited to sheet-metal design and nesting problems, while for complex assemblies the interference checking ability of 3D solid modelling is essential. A prototype design rule checker for sheet-metal, which extracts feature information from the CADDS 2D CAD system, is presented by Glovin and Peters (1987). Another example of the use of a 2D CAD representation as a basis for a DFM system is presented by Terry *et al.* (1990). Here, a 2D CAD system is used to produce a tool model from user specifications, which is then checked for design faults. The computer-aided DFM/A systems reported in research tend to be specific to particular domains. The importance of the generic system is that it aims to provide a single environment for DFM/A analysis over a number of domains, such as electronic and mechanical assembly, and machining.

Clearly the earlier in the design process that DFM/A can be implemented the better. Many companies operate a "gated" or phased product development review process, with particular criteria to be met before the end of each phase is agreed. Therefore in reality, companies may also desire different levels of DFM/A capability corresponding to the different design phases. Companies may have different names and particular definitions for these phases but in general designers are dealing with varying levels of functional, qualitative and quantitative information (Figure 1.6). The PCB design process for example progresses from a specification, through a functional circuit diagram, developed with a specialist electronics design system such as PSpice, to a full board layout including routing and layer information. Obviously to implement DFM/A at all of these phases would require greatly differing types of interfaces and analytical capability, and the use of a central common product data model would be extremely useful.

For CAD product information (normally product structure and geometry - additional functional information must be obtained elsewhere) to be analysed by DFM/A systems, it must be interpreted with respect to the actual manufacturing or assembly domain. The knowledge applied by a DFM system is generally elicited or acquired from manufacturing personnel, while the interfacing of CAD and DFM/A systems requires the translation of information between different domains and vocabularies, as the knowledge acquired will be domain-specific. For example, whereas a designer may see a hole as being a functional entity, a machining process planner may see it as a material removal process.

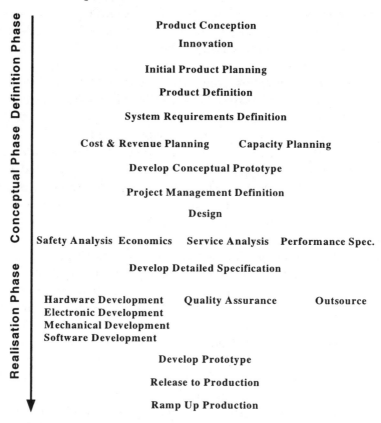

Figure 1.6 *Sample Product Development Process*

The knowledge acquired may also refer to current regulations or limits of the domain X. For example, the knowledge governing Design For Machining refers to a mixture of good practice and available machining process capability. Similarly, Design For Environment knowledge may refer to current accepted practices as well as prevailing legal constraints.

Whereas this translation, say from design features to manufacturing features, may be performed separately to the DFM/A analysis, it is often combined with the DFM/A rules, such that the recognition of a particular design situation implies a particular DFM/A problem. DFA guidelines issued to a PCB layout designer may state rules in the following manner:

components should be at least 1.0 cm from the board edge.

This does not tell the designer the origin or meaning of the rule, or whether .99cm might also be acceptable. In the mind of the manufacturing engineer the actual knowledge may exist as:

components should be more than the minimum conveyor width from the board edge. Conveyor width for machine A is 1.0 cm and for machine B it is 0.9 cm.

Thus the content of the DFA rule should where possible be directly related to the process plan and equipment model for a particular product for greatest flexibility. In general for DFM/A knowledge to be most accurate and applicable to the widest variety of situations, it must act as the link between the design domain using the prevailing constraints from domain X. The use of a common product model would also enhance the ability to trace design decisions during the design process.

One issue which must be addressed in interfacing CAD and DFM/A systems is how they will interact with the designer. The usual distinctions are between on-line and off-line analysis. The DFM/A system may be on-line with the CAD system, with analysis triggered by certain actions of the designer, or it may be on-line waiting until the designer requests an analysis. They may share the same product model or maintain separate models. The DFM/A application may operate using macros written in the CAD system development language or the CAD system may form part of the DFM/A tool. For example, the ACIS modelling software (produced by Spatial Technology Inc.) is currently being used by some research institutes as a product modelling tool embedded in applications such as DFM/A (DEFMAT, 1995). This solves the problems of communications and of data coherence management, but precludes genericity with respect to CAD systems.

In off-line analysis, the CAD data is input to the DFM/A system using a neutral format file, with no direct communication with or feedback to the CAD system. Whether on-line or off-line analysis is used, both require that a standard for representation and exchange of product data between applications is established. As CAD systems' internal representation cannot fully support DFM/A, other standards are needed. In the late 1970's, in answer to the need for data transfer between CAD systems as well as between CAD and CAM systems, IGES (Initial Graphical Exchange Standard) was jointly developed by Boeing, General Electric and the United States Air Force (Liu and Fischer, 1993). IGES functions well for 2D drawing applications (for which it's data structures were originally intended), but is weak in Solid Modelling. The Autodesk proprietary standard, DXF, is also widely accepted and supported for 2D applications. Information such as features are not represented. This led to the setting up of the PDES (Product Data Exchange Specification) initiative in 1984, with the objective of defining all the information needed to design, manufacture and support a product. This initiative is now part of STEP (Standard for the Exchange of Product Model Data), which is an international movement to produce a standard which will define an external representation of a product model for all the phases of the product life-cycle (Scholz-Reiter, 1992). Within the European Union, the Esprit project 322 (CAD*Interfaces) has developed a neutral format in co-operation with STEP which allows communications between 12 different CAD systems. According to Cutting-Decelle and Dubois (1994), STEP refers to two main notions: product modelling (the concept of product data), and the exchange of product data generated by product modelling. As illustrated in Figure 1.7, the different categories of product life-cycle data defined under STEP assist in realising the concept of Concurrent Engineering using an integrated product life-cycle model, supporting various perspectives on the product (or project), as well as exchange of information between design support tools.

Both Cutting-Decelle and McKay *et al.* (1994) contend that STEP is currently in danger of being torn between conflicting objectives, with ensuing dissipation of energy and failure to meet even its original objective of data exchange. This seems to be caused by a desire to address not only the problem of data exchange but also the problem of an integrated product representation that covers all phases of a product life-cycle, which could be used for information sharing. The main outputs so far from STEP have been the definition of a number of application protocols and the Express language. Application protocols define the form and content of a block of data that is to be exchanged such that software products purporting to conform to this standard may be tested.

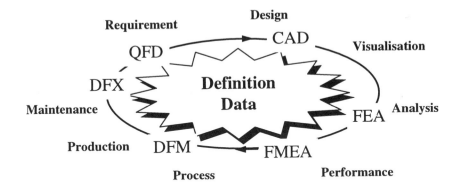

Figure 1.7 *Categorisation of Product Data*

Grabowski *et al.* (1994) present an example of an application protocol for the mechanical design process for the automotive industry (AP 214), currently under development, which was initiated by the German STEP centre and the VDAI (the German automobile standards association), who were also very active in developing subsets of IGES to address the needs of the German automobile industry. An application protocol describes the scope, context and information requirements for the application and specifies those parts of STEP which satisfy those requirements. These requirements are formalised in AP 214 using Express and the IDEF and NIAM modelling languages.

The most widely used part of STEP is currently the Express modelling language (Express, 1990) which consists of the object-oriented meta-language which is used for STEP information modelling and Express-G, which is the graphical representation of Express (see Appendix A for a description of the Express language) and is somewhat similar to IDEF1X. Using the Express language capability, it is possible to not only capture attributes of entities but also store within entities, rules concerning attributes and references to operations on those attributes. Thus there is the possibility of exploiting the full capabilities of object-oriented programming using Express. For example, Liu and Fischer (1993) reports on the use of Express to define C++ classes for manufacturing features which may be stored in an object-oriented database. A number of public-domain and commercial tools are available which perform a variety of Express-related tasks such as syntax checking, graphical editing, translating to databases, generating C++ classes from Express, etc.. For example the ROSE system (Spooner, 1994) is being developed by a US research team to develop a STEP/PDES compliant product database, where Express entities are converted into persistent C++ objects. Express will be used as the main data modelling language in this book.

In order that DFM/A may be most easily applied to products, some preliminary work has usually been necessary to translate the purely geometrical product CAD model into known features recognisable within the domain X. There has thus been a lot of emphasis in recent years on feature recognition (also called feature extraction) (Bronsvoort and Jansen, 1993) (Goldbogen *et al.*, 1988), (Henderson and Chang, 1987, 1988), (Joshi and Chang, 1990) with the vast bulk of the research focusing on feature recognition for machining process planning (Chang, 1990), (Honda *et al.*, 1993), (Zhang and Alting, 1993). This problem of recognition of features such as slots, holes, grooves etc., presents the greatest obstacle to the implementation of integrated CAD/CAPP or CAD/DFM systems. It is partially solved by feature-based design systems, but is further complicated by intersection between features. Delbressine (1989) uses a variation on the feature-based design approach, using manufacturable objects and transformations to define products and processes. The aim here is to force the designer to work only with manufacturable objects, ensuring manufacturability and speeding up the process planning activity. In these respects it suffers the same drawbacks as the other feature-based design methods, as well as over-constraining the designer. Many different methods of feature recognition have been investigated, but none is perfect. Joshi and Chang (1990) present a good survey of the different recognition methods. While feature-based design seems to offer the solution to the problem of feature recognition for process planning purposes, feature recognition will still probably be necessary in the future for the following reasons:

1. Feature based design will not be capable of producing all of the geometry necessary for the definition of a product, and intersections between features are inevitable.
2. While a certain set of features may be found adequate for designing for a particular process, if the process is changed the feature

3. definitions must inevitably change also. Feature recognition might be needed to convert features from one set of processes to another.
4. Feature based design is restrictive, and it would not be possible to create a big enough library of features for all future designers needs. Even if it were, the task of searching for the correct feature would be prohibitively slow.

However, to avoid concentrating exclusively on the problems of feature recognition, in this book it is presumed that the full set of design features needed for DFM/A analysis is available to the designer through the CAD system.

1.8 Conclusions

An overview of current research in DFM/A, and in particular DFM/A, has been presented. Existing methods, using spreadsheet type analysis, are not adequate for the needs of many industries. What is needed is a knowledge-based AI approach, where the DFM/A knowledge may be customised by enterprises, providing automatic evaluation of the designs. DFM/A constraints however, must be modelled concurrently within the design process to allow designers from an early stage to assess the implications of design decisions on all aspects of the product, from customer requirements to environmental constraints. To fully support the DFM/A negotiation process between different domains, it is necessary to have the support of a complete life-cycle product model. Some progress has been made in this area in STEP, and in Chapter 4 we explore in more detail the product modelling requirements for Concurrent Engineering, and in particular, for advanced DFM/A systems.

2. Design for Manufacture and Assembly Methodologies

2.1 Introduction

The generally accepted model of the design process is as something which outputs a solution based on given requirements, with inputs from the different life-cycle phases (Figure 2.1).

Obviously, the requirements for each industry, perhaps even each company, will differ in the way they divide the design process. However the following phases are common to most product design methods:

1. analysis and requirements definition
2. functional design
3. conceptual design, consisting of:
   a) product structure design (determination of main components and subassemblies)
   b) design sizing (specifying component relationships and main parameters)
   c) identification of appropriate available production technologies
4. detailed design, resulting in the product's total form.

The first two design phases are concerned with discovering or formulating the problem, and with expressing it in such a way as to describe the ideal solution. The conceptual design phase is then concerned with finding an acceptable embodiment of that solution. At the beginning of the conceptual design phase, the designer must begin to narrow the range of possible solutions, for example with respect to the range of production technologies available, or the overall arrangement of functionalities within the product. Thus the designer is dealing with a mixture of constraints, coming from market, customer and production areas, varying from the qualitative to the quantitative, which assists him in narrowing the set of possible solutions down to the chosen one, which must then be detailed further. From a design automation viewpoint, we see a variety of techniques utilised in this process (Figure 2.2). In this chapter we look at ways in which manufacturing and assembly constraints can be integrated into the design process, and special requirements for different industry types.

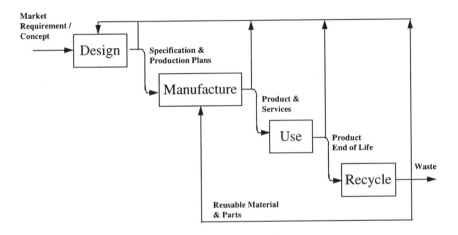

Figure 2.1 *The Product Life-Cycle (Simplified)*

2.2 Moves towards a total design environment

To allow full integration of manufacturing constraints into the design process and allow concurrent development of the different product life aspects, a coherent, comprehensive product and process model is required. As stated by Krause *et al.* (1993a) in their review of the state-of-the-art in product modelling *"product modelling is the key factor in determining the success of various product development strategies and industrial competitiveness in the future"*. To date standards such as IGES, SET and DXF have come into being to allow communication of mainly CAD data between applications. However these require pre- and post-processors and do not support full product life-cycle data. A product model should be able to support complete life-cycle aspects. The most important initiative towards developing such a product model is the development of the ISO Standard 10303 "Standard for the Exchange of Product Model Data" (STEP).

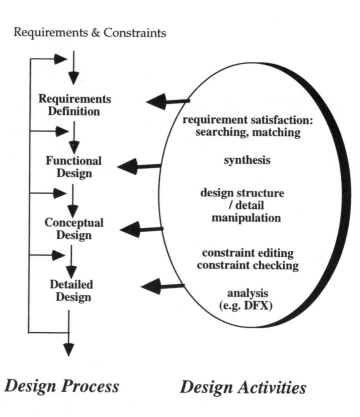

Design Process   Design Activities

Figure 2.2 *Basic design activities*

The goal of the STEP initiative is an integrated approach to document all the relevant data about a product. This involves everything from geometry and topology descriptions to production and maintenance aspects, combined into a single, consistent logical schema (Hordvik and Oehlmann, 1992). Thus, going beyond the merely geometrical shape descriptions normally found in CAD systems, the product model concept is generally considered essential to CE in an automated design environment. As the STEP initiative has yet to produce such a standard, most of the developments in intelligent CAD systems mentioned above are being done using different proprietary product data representations.

The STEP/Express modelling formalism (most often seen as Express-Graphic data structure diagrams) has been the most visible output of STEP, so far adopted mostly by the research community. However, while the rules of the Express language itself are fairly well defined, the STEP product model has yet to be produced. Some advanced work has been done by projects such as BRITE DEFMAT (Molloy, 1992), Esprit IMPPACT (Meier, 1991) and CIM PLATO (Seliger *et al.,* 1992) in the use of Express in modelling integrated product/process models. A complete product model approach should support several aspects of product development. Three main points listed by Hordvik and Oehlmann (1992) for attention in product modelling are:

1. product structure (e.g. hierarchy), including conceptual aspects such as function
2. version change and management
3. varying levels of concretisation (and detail) of product constituents.

Other aspects which would need support in a complete product model are:

4. multiple perspectives by different CE team members
5. multiple access control
6. ability to capture design phases from customer requirements to production requirements
7. linking of product model to life-cycle phases (e.g. design parameters to process parameters) for early design decision support
8. incorporate attributes such as targets, constraints, tolerances and relationships on design objects
9. varying levels of certainty and quantitativity
10. cost factors
11. capture of evolving product model and design decisions
12. support for multi-domain products such as mechatronics products with interfaces between different component types (software, mechanical, electronic).

A review of the current basic requirements and technological challenges of product modelling is contained in Krause *et al.* (1993b).

2.3   Tools for total design

Traditionally, the sequential engineering (*over the wall*) approach (Figure 2.3) to product development consisted of a series of sequential steps from product concept right through to product maintenance, wherein the information flow was uni-directional. In this type of environment, the amount of interaction between the different development phases was limited largely to adjacent phases. These iterations between phases often involve long delays with various translations of data formats and vocabulary. For example, a designer will use a particular set of symbols and operations in his work which are composed of basic geometric entities in a CAD system.

This representation must be interpreted by the process planner, effectively translating to a different set of symbols and operations. That is, the sequential development process is set up in such a way that direct dialogue (both machine-based and human) and information exchange between, say, conceptual design and manufacture is severely limited by the absence of the appropriate common culture and mechanisms.

Effectively, sequential engineering considered the specific purpose of the design department to be the design, development and redesign of a product. Succeeding this, the manufacturing phase aimed at producing these products. The sequential engineering philosophies fostered inter-departmental rivalry and closed communication, with the result being high product failure. Much work, for example has gone on in the area of detailed design-process planning integration alone, with feature-based design and DFM/A acting as integration and communication mechanisms.

To fully exploit, however, the advantages of CE, mechanisms should exist to allow all phases of the product life to integrate and communicate effectively.

2.3.1   Quality Function Deployment

The tracing of relationships between product and production parameters is a vital aspect of Concurrent Engineering design, which should be performed for all phases of the product life-cycle. One methodology which supports this aspect of CE and is best suited for team use is Quality Function Deployment (QFD). QFD originated in Japan during the 1970's as a systematic technique for identifying those product features which contribute strongly to product quality, and thus where engineering effort is needed. QFD is based on a matrix approach to design, mapping the requirements (starting with customer requirements) onto the means of achieving them. Therefore a series of charts may be developed which map the relationships between customer requirements, engineering characteristics, right through to production planning. An example of a QFD chart is shown in Figure 2.4.

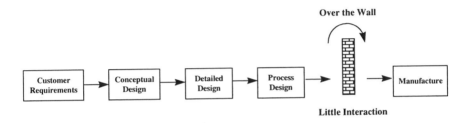

Figure 2.3 *Sequential Engineering*

The matrix rows represent the customer requirements, and the columns the engineering characteristics. Thus, the relationship between customer requirements and engineering characteristics can be identified in this matrix. A "cascade" of charts can be created, dealing with the different aspects of the product life-cycle, at the design stage (Figure 2.5). As they can be worked on concurrently, tracing cause and effect between the different stages, QFD is an effective team tool for product life-cycle design.

QFD is not as such a product development methodology. However it is concerned with the coherent mapping of customer requirements through the product and process development process. Through the matrices, mappings or relationships between different domains are established, as well as targets for the attributes. What QFD mainly does is help identify critical characteristics, rank characteristics, establish relationships between characteristics and establish targets for characteristics. It may also help identify characteristics which are redundant. There is also clearly scope in QFD for identifying cases of functional coupling between design parameters in a more readable format.

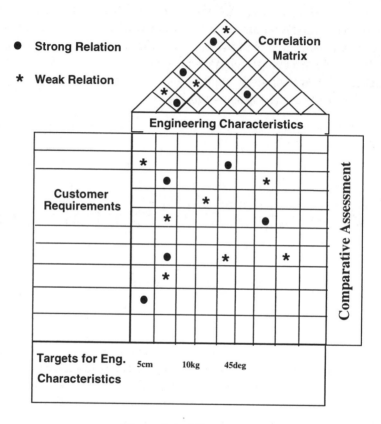

Figure 2.4 *QFD Chart*

Liner (1992), in an evaluation of first experiences in using QFD in a large company, recommends its use from the beginning of development projects. Thackeray and Treeck (1990) and Egenton and O'Sullivan (1992) both demonstrate the use of QFD in software product development. Most requirements analysis tools (such as IDEF) use data and control flow diagrams to specify the requirements of software products. However QFD offers the following advantages over these methods:

1. it shows mappings between requirements and components
2. it shows potential requirements conflicts
3. it shows the positive or negative impact of design elements on requirements
4. it shows the positive or negative impact of design elements on other design elements.

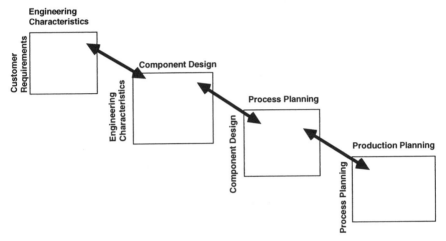

Figure 2.5 *QFD cascade*

There are some potential dangers in adopting QFD for CE teams. In a study of communication patterns among two sets of product development groups using QFD and phase-review in the US, Griffin (1992) found that QFD did encourage teams to become more integrated and co-operative. However they also seemed to be more self-sufficient, looking for solutions within the team rather than searching for project-related information elsewhere in the organisation. In another study of 35 US companies Griffin and Hauser (1992) conclude that the way in which QFD is implemented may greatly impact the companies' abilities to see any measurable benefits. A high level of commitment at all levels is required. QFD is seen as just one technique among a number that are available for developing new products. The basic QFD matrix (Figure 2.6), relating requirements to solutions via a relationships matrix, reflects the activities of the design process and can thus be applied to most design problems.

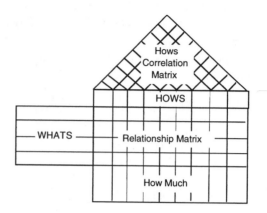

Figure 2.6 *Basic QFD matrix*

Further, it should be possible to use the QFD principle in many ways, for example adding recycling planning to customer requirements or as limitations on certain engineering characteristics, or even treating environmental concerns as another "customer" (Figure 2.7). This would mean that changes in component design or even production planning could be traced to environmental concerns as well as customer requirements. Krause *et al.* (1993b) suggest that *"...the potential of a QFD application is exhausted only by an integrated application using CAD methods based on integrated product models"*.

In both Pugh's Total Design methodology (Pugh, 1991) and in Olesen (1992) the QFD charts are shown as being performed in sequence, one at each stage of the design process, whereas this need not be the case. QFD type chart cascades could be started right from the beginning of the design process, with relationships between the different product life phases traceable through the charts. What is important is that the network of relationships in the QFD charts is available throughout the design process, and is accessible by all design tools, to maintain coherence.

The potential complexity of the QFD method, when dealing with a reasonably complex product, can be daunting. Perry (1992) shows that in a project dealing with 128 customer requirements, correlated with 65 major product features, a matrix of over 8,000 relations results. Thus manual methods in such problems are tedious, discouraging the implementation of QFD and the reuse of previous analyses. Fortunately a number of computerised QFD packages are now on the market, such as QFD Designer produced by the QualiSoft Corporation, and QFD Capture, produced by ITI. The QFD software currently available makes QFD accessible and reusable, but does not provide direct links to the manufacturing process and tools such as DFM/A - strongly advocated by O'Connor *et al.* (1992) in a paper reporting on a prototype QFD software tool.

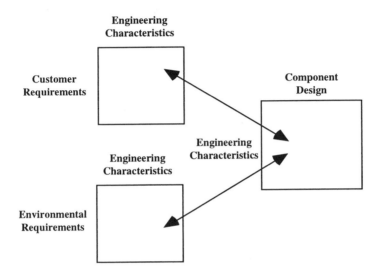

Figure 2.7 *Further QFD possibilities*

Another methodology used by a large number of companies in the area of product / process defect prevention and quality assurance is Failure Mode and Effects Analysis (FMEA) (McMahon and Browne, 1993), which is a preventative quality method involving the examination of design and production drafts for possible weak points in the planning stage.

2.3.2 Failure Modes and Effects Analysis (FMEA)

First used in the development of flight control systems in the early 1950's, FMEA has become increasingly popular. For many companies, FMEA forms an important part of their CE philosophy. FMEA is used particularly in relation to the evaluation of product reliability and safety in use. By tracing the effects of component failures through sub-systems to system failure modes, the impact of possible failure modes at all levels can be assessed, and preventive or corrective measures taken. The primary aim is to benefit from knowledge and experience to reduce the probabilities of certain failures occurring / recurring.

Two main applications of FMEA principles may be seen in product and process design. While product FMEA is reliability assurance, process FMEA is planning for quality in production. Process FMEA begins by assuming a failure and focuses on the modes that could cause that failure to occur. Product FMEA uses an inductive approach by recognising common failure modes and examining the effects of those failures on the entire system, or sub-system, depending on the scope of the FMEA.

FMEA is a data intensive method, and to be performed successfully requires detailed and time-consuming product performance and process analysis. It must also be well documented in order to provide a clear well-related hierarchy of data. According to Vincoli (1993), to facilitate the construction of useful FMEA's, some or all of the following information should be gathered:

1. Design Drawings
2. System Schematics
3. Functional Diagrams
4. Previous Analytical Data
5. System Descriptions
6. Previous Experiences
7. Relevant Component Specifications
8. Preliminary Hazard List
9. Preliminary Hazard Analysis
10. Operating and Support Hazard Analysis
11. Other System Analyses previously performed
12. Customer Feedback
13. Process Control.

Once this information has been gathered the nature and scope of the FMEA must be decided. This will ensure the FMEA analysis has the necessary focus to achieve meaningful results. A number of different worksheets are in use, usually differing from company to company. A typical worksheet is shown in Figure 2.8. The typical FMEA procedure described below is an amalgam of several procedures currently in use:

1. Define the system, it's purpose, performance requirements and operating environment.
2. Divide the system into sub-systems; break the sub-systems into their main functional blocks; describe their hardware and software components an the interface between them.
3. Using the FMEA worksheet, prepare a component list at the level analysis is to be performed (e.g. sub-system level). Define the function of each component.
4. Define all potential failure modes to be evaluated at the lowest level of assembly.
5. Define the effect of each failure mode on the immediate component function.
6. Detail preventive or corrective measures to eliminate or control the effects of the failure.
7. Rank the failures in terms of severity. For example, on a scale of 1 (minor, unlikely to cause system failure or jeopardise operator safety) to 10 (a failure causing immediate system failure, with potentially harmful effects).
8. For each failure mode, indicate how it will be detected, and any design measures that have been taken to compensate.
9. Define (if data is available) or compute failure rates. For example, a score of 1 indicates that the failure modes risk is insignificant, whereas 10 would indicate it is likely to occur regularly.
10. Based on currently available quality control mechanisms, estimate or provide data on likelihood of failure mode detection. Again these can be score from 1 to 10.
11. The product of the failure mode severity, failure rate and detection figures is termed the Risk Priority Number (RPN). This figure (essentially a qualitative means of prioritising failure modes), can be used to rank the failure modes. Thus high RPN's would require most immediate attention.
12. Finally define changes to the current design or operating procedures to reduce the overall RPN figure. This is normally achieved by setting out an action plan, with new targets for one or all of the failure rates, severity and detection figures.

Vital to the long term benefits of FMEA is the regular review and updating of previous analyses. Essentially, if the action plans have been successfully executed, the targets set should become the new actual failure rate, severity and detection figures. Thus the FMEA cycle can be maintained until all failure modes requiring attention (with an RPN over an accepted level) have been addressed.

As well as the obvious benefits of improved product and process design, the following benefits will accrue from properly executed FMEA's:

1. Through the analysis process, a repository of product and process knowledge is gathered.
2. The analysis should lead to improved CE teamwork.
3. High risk areas can be identified and actioned.
4. Diagnostic procedures should be improved.
5. Failure rate and probability data gathered can be used to develop inspection and maintenance schedules.

Due to the tedium of repeating FMEA analyses, and transferring data from chart to chart, as well as the difficulty of handling large system FMEA's, a large number of commercially available FMEA systems are now on the market. The following drawbacks however are still associated with the current forms/spreadsheet type FMEA and QFD software which:

1. cannot exchange information with other software tools
2. cannot make use of current quality information in a dynamic fashion
3. cannot be generated from the product design
4. cannot automatically make use of existing expertise to propose solutions / actions
5. do not store data such that the inherent relationships are retained
6. do not reuse previous knowledge (as opposed to data)
7. do not support or promote teamwork.

In the ideal CE design situation, work on the design of all aspects of the product's life should proceed concurrently. However, hierarchical decomposition of the design problem, coupled with the ability to move between design domains would normally be required. Some work has indeed already been done by Wallace and Suh (1993) on a generic shell for interactive design tools based on the axiomatic design concept. Their

Product: A/C Adapter
Part No.: 101CB
Revision: 3.1

Failure Modes & Effects Analysis
(Process FMEA)

FMEA Team: Quality 1
Date: 27/10/1996
Rev. No: 3

Process Description	Potential Failure Mode	Potential Effects of Failure	Severity	Potential Causes of Failure	Occurrence	Current Controls	Detection	RPN	Recommended Action	Resp. and Compl. Date	Action Results				
											Actions Taken	SEV	OCC	DET	RPN
Leadframe Preparation	Damaged Leadframe	Cannot Insert Into Tooling	6	Wrong Fixture Used	2	Identify Fixture	3	36	Fixture Labelling	Process Technician	New Labels Developed	6	2	2	24
		Cannot Place On PCB		Fixture Setup Wrong	3	Training or Design for Access	4	72	Special Training	Team Leader	Training Session Given	6	2	3	36
	Incorrect Preparation	Sequenced Incorrectly	7	Wrong Fixture	2	Identify Fixture	5	70	Fixture Labelling	Process Technician	New Labels Developed	7	1	3	21
		Trimmed Incorrectly		Tools Worn	1	Maintenance Programme	3	21	Maintenance Programme Review	Maintenance Supervisor	Maintenance Schedules Revised	7	1	2	14
	Wrong Parts Used	Parts Confused	8	Missing ID	3	Part ID's on BOM	2	48	Storage Bins	Process Technician	New Storage Bins Introduced	8	2	1	16
		Incorrect Leadframe Used		Operator Error	2	Training Programme	3	48	Special Training	Team Leader	Training Session Given	8	1	2	16

Figure 2.8 *Sample FMEA Worksheet*

aim is to allow designers "....*in a seamless manner to view designs from alternative life-cycle perspectives*". Using the information-content method of determining design efficiency, they relate the customer domain (product attributes) to the functional domain (design goals), to recommend goals and strategies before beginning design. They make the point that qualitative approaches such as QFD are not as well suited to computerisation.

It is true that qualitative reasoning is difficult to implement successfully, but conversely, as one moves towards conceptual design, the level of certainty decreases, and hence the qualitativity of information increases. Therefore qualitative reasoning must be included in future design systems which aim to cover the conceptual stage. In fact, work has been done by people such as Jackson (1991), Jackson and Browne (1992), in qualitative modelling to support strategy determination, and by Molloy (1995) in Concurrent Engineering.

Techniques such as QFD provide a mechanism which allows a mix of quantitative and qualitative reasoning to support goal and strategy determination. For large projects however the QFD charting technique becomes very unwieldy and will result in criss-crossing chart cascades for different subassemblies.

Fundamental to the conceptual design stage is functional design, which is not supported by most current commercial CAD systems. Some work has been done by Mantyla (1990), Mantyla *et al.* (1994), Sivard *et al.* (1993) and Tomiyama *et al.* (1993) in the development of design systems which support constraint satisfaction and functional decomposition. All three use object representation operating on the product design level. Frame-based constraint networks are also being used to produce CE design advisors.

Due to the large amount of potential information which will have a bearing on the life of a product and the complexity of the product itself, tools must be provided to the individual designer and CE team to support the product development process. A large amount of research is ongoing in Japan (Kuo and Hsu, 1990), into Intelligent CAD (ICAD - not to be confused with the commercial CAD system of the same name) to assist designers and contribute to the integration of CAD and CAM. Ishii *et al.* (1993b) present an ICAD system based on a knowledge base containing knowledge about physical phenomena based on qualitative physics.

Using the knowledge base, physical phenomena can be derived from the structure of the object, assisting the designer in building the structure of an object which exhibits the desired behaviour. Feru *et al.* (1993) have developed a system which allows a designer to create a product from a functional decomposition approach, taking into account manufacturing considerations, based on an object-oriented central data structure.

2.4   Design for Manufacturing and Assembly Principles

In this section we demonstrate the DFM and DFA requirements of different industry types. It is not meant as an exhaustive compendium of manufacturability guidelines, but to illustrate the requirements for integrating DFM/DFA into different design/manufacturing regimes. In all real-life cases, what is required is an informed negotiation process between the requirements of design and manufacturing to produce a viable solution.

2.4.1   Mechanical Assembly

Very often in mechanical assembly, as in most other industries, the need to satisfy increased customer demand for quality product, at lower cost in shorter time, is leading to increased automation of previously manual assembly processes. The inherent flexibility of manual systems, coupled with greater profit margins in the past, often led to each design engineer having preferred methods in design and making parts. This is one reason for special solutions. For example, each engineer may have used different ranges of screws, purely for historical reasons. In automated assembly of course, such approaches quickly lead to complications in handling and feeding equipment design.

  A problem, particularly in environments becoming automated, is that the design engineer checks the functionality of the product but not always the design for assembly, in particular with regard to the available process and equipment. Therefore, for the manufacturing of a new product, often new tools, parts and equipment have to be purchased. This results in higher costs and a large effort of time to select, install and test the new hardware components. So the designer must be encouraged to adopt a DFM/A approach.

DFA Application in the Mechanical Product Development Cycle

The main objective of design for assembly is to reduce assembly costs. However, the effect of product design on the whole manufacturing process and product life-cycle should be considered. In general terms we can apply DFA at all stages of product design. In the mechanical engineering design domain, we may broadly speak of the "concept" and "detailed" design stages. We may also broadly categorise DFM/A rules in terms of "general", as in rules of thumb, "process- or industry-specific", such as relating to mechanical assembly, and "company-specific", where particular production parameters, or preferred practice, dictate the application of DFM/A rules in specific cases.

Where should DFA be applied in the product life-cycle? This is a common question in most industry sectors. There are arguments that DFA should be applied at the conceptual stage, other arguments suggest that it should be applied during the detailed design stage. The approach taken in this book follows the school of thought that DFA should be applied as early as possible in the product life-cycle because of the assembly cost commitments which are incurred at the early stages of the product design. DFA should also be applied throughout the product life-cycle to guarantee that designs comply with life-cycle requirements. This approach refines the application of DFA so that certain rules are applied earlier in the product life-cycle.

A large number of rules for design for assembly have being developed and tested by various people working in the DFA field. Most of these rules are domain-specific. However, in developing a generic methodology and architecture for DFA systems, the following DFA principles should be considered.

2.4.2   General DFA Principles

1. Minimise number of parts
If the number of parts were to be reduced, this would reduce the number of joining operations required and hence, hopefully, reduce the assembly cycle time. The main method of reducing the number of parts is by combining two existing parts. Of course it could be argued that combining two existing parts may increase the complexity of the new part. Therefore, this part may be difficult to handle, etc. and the cycle

time may, instead of being reduced, be increased. Consideration should be given to the other desirable assemblability characteristics when applying this principle. It should be applied throughout the product life-cycle particularly during the conceptual and detail design stages.

2. Design for Ease of Handling

Good handling features are essential if automatic assembly is to succeed. In order to have good handling (gripping) features it is imperative that components display the following attributes:

1. provide gripping faces
2. geometry - components should be of a regular shape
3. stiffness - avoid soft brittle materials which may bend during handling and assembly.

If component are not designed to cater for the above then there exists a need to design and produce fixtures for these components. This increases assembly cycle time because of the set-up times required for placing the components in the fixtures. It also adds to the cost of the assembly and reduces the design flexibility. If components don't have good handling features then assembly becomes difficult; be it automatic or manual assembly. This applies to most product domains. This principle should be applied during the detail design stage.

3. Design for Ease of Insertion

The size of product components can pose real insertion problems. The robotic equipment needed for inserting components in the automated assembly domain has to be very accurate and process a high repeatability factor, otherwise insertion would become virtually impossible. To design good insertion into the components which need this characteristic, the designer can take the usual approach of having tight tolerances and chamfers on both the inserted component and the other component. This principle should be applied during the detail design stage.

4. Standardise parts

The system debug and set-up times for automated assembly processes may be very high. This results in long product to market time and high assembly cost. Also the process equipment design is complex and expensive. In addition to these considerations, the overhead for a company in terms of maintaining part inventories can be quite considerable. For example, Alcatel reported several years ago that the calculated overhead cost to add one part to inventory was of the order of $25,000, excluding purchase price. In order to reduce overall assembly time and cost, standard parts should be used if possible. This would lead to the following benefits:

1. minimum changes to the assembly process set-up
2. reduced inventory problems
3. reduced planning and scheduling
4. high probability of the right design first time.

This principle should be applied during the conceptual and detail design stages.

5. Design for current process capabilities

It is essential to take process capabilities into consideration at the design stage in order to assess how the design of the component can be constrained by the process. This should reduce the chances of designing components that cannot be assembled using the current assembly process capabilities. This principle should be applied during the detail design stage.

6. Maintain awareness of alternative process capabilities

Reasons may exist which dictate that designs cannot conform to existing in-house process constraints, for example product functionality. In this situation the most optimum solution must be chosen. The reason for this principle is to ensure that designers have an awareness of process capabilities which could be available if required and justified. A costing mechanism should be provided to allow the designer to gauge the financial implications of designing outside current process capabilities. Such a mechanism would act as an advisor allowing designers to determine whether it would be more economical to employ technology for assembly or to stick with existing process capabilities. This principle should be applied during the detail design stage. It should be noted that if standard parts are used then principles 5 and 6 are not applicable.

Product design must accommodate multiple life-cycle values. So product design and production planning should be integrated into one common activity. This is a basis for a common consideration of the product and the assembly process. It leads to "process driven design", as required by industry. Figure 2.9 illustrates how such principles, properly applied, can be beneficial by addressing them at the earliest design stages.

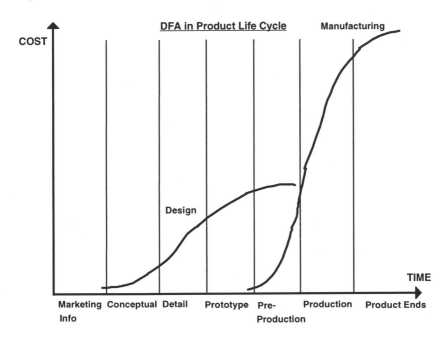

Figure 2.9 *DFA in Product Life-Cycle*

Usually the above principles are restated as General DFA Rules, which may vary depending on the design/manufacturing regime in question. Two sets of general DFA guidelines, applied to the mechanical and electro-mechanical regimes are listed below.

2.4.3 *General mechanical DFA guidelines*

1. Reduce the number of assembly processes
2. Reduce the number of single parts
3. Use integral parts
4. Use symmetric parts, avoid pseudo-symmetric parts
5. Ensure accessibility to all fixings and joining points
6. Use standard tools and process materials

7. Watch for free handling space
8. Make sure of pre-positioning
9. Ensure easy access for testing
10. Provide modular component structure
11. Differentiate variants at the end of process
12. Check material compatibility
13. Prefer flow parts instead of single parts

2.4.4   General electro-mechanical DFA guidelines

1. Minimise parts for radio-shielding
2. Use only one base-part
3. High frequency shields have to be ensured
4. Weight and dimensions have to be small
5. Parts have to be shock-resistant
6. Minimise joining directions
7. Minimise use of flexible parts
8. Minimise the number of different screw diameters
9. Use only torque-screws for automation
10. Design to allow functional subassemblies to be tested independently
11. Use gripping faces with the same dimension
12. Use bearing holes in all PCBs at the same location.

2.4.5   Design for Manual Assembly

If we are to view DFA as facilitating the assembly process through design, then Design for Manual Assembly must be of primary concern. In particular, handling of parts may account for almost 80% of total assembly time. While the same principles may apply as for automated assembly, particular consideration must be given to the capacities of human operators as integral parts of the assembly process. The following checklist highlights some of these issues:

1. Avoid slight asymmetries
2. Minimise part orientations
3. Eliminate sharp corners
4. Chamfer lead-ins

5. Minimise part weight
6. Eliminate fasteners
7. Minimise tool requirements
8. Eliminate adjustments (through tolerancing)
9. Avoid moving subassemblies.

If we are to view the disassembly process, whether at the end of product life, or as part of routine maintenance, as to some extent the inverse of the assembly process, then consideration should be given to issues such as accessibility. The types of operation to be performed, e.g. requiring the use of tools to remove parts requiring replacement or servicing, should also be considered (Figure 2.10).

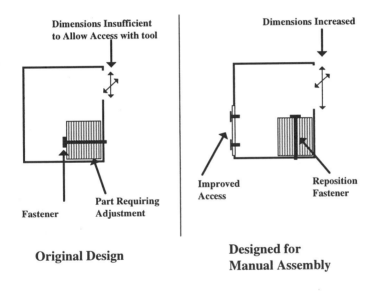

Figure 2.10 *Design For Manual Assembly Example*

Some of the problems caused by poor accessibility during manual assembly are:

1. Worker Fatigue
2. Decreased Quality
3. Decreased Productivity
4. Increased Manufacturing Time and Cost

5. Increased Safety Risks
6. Increased Training Time.

Recent work in DFA has concentrated on automatic assembly, although some older work was done on the evaluation of products for manual assembly for example by Fearing (1982), and Sturges and Wright (1989). The actual kinematic issues involved have been studied by Salisbury and Craig (1982). Sturges developed a method for the quantification of manual dexterity based on human motor capacity, which is applied in the form of a slide-rule for the manual evaluation of assemblies. He also presents two useful sets of design checklists for parts and assemblies collected from industrial experience, which consider the assembly process from a dexterity point of view rather than the from/fit/function of the parts. However to date this interesting work does not seem to have been incorporated into more recent research into DFM/A expert systems. While the principles of Design For Manual Assembly are incorporated into methodologies such as Boothroyd-Dewhurst or Lucas, the more analytical approaches to the problem have not been apparent in current DFM/A solutions.

2.4.6   Electronics Assembly

In the increasingly complex and cost-conscious world of electronics design and manufacturing, Design for Manufacture remains a high priority. In fact Holden (1995) reports that in the electronic design process 60% of the manufacturing costs are determined early in the design process, with only 35% of the design costs spent. Much work remains to be done in this area, even within apparently sophisticated multinationals, with geographically widespread design and manufacturing functions. A number of issues must be addressed for CE to succeed in the competitive and geographically diverse electronics industry:

1. Extreme pressure exists in design sites to release product, from a time to market and cost perspective. Process yield and ability to manufacture are not always seen as major drivers for designers.

2. Design Engineering are often not familiar enough with manufacturing processes. Designers need to be made aware of the problems caused in manufacturing by incorrect definition of items, e.g. insertion codes, rotations, pads, pitch, and x'y co-ordinates.

3. Where products are manufactured in several locations, design sites tend to focus on the prime manufacturing plant's process requirements, from a module layout perspective. This can cause problems in a secondary plant with a different set of equipment. For example one site may use Fuji placement equipment, whereas another may use Siemens. Design sites usually concentrate on running producibility checks at design completion, with the prime plant's process in mind.

4. DFM input from the manufacturing plants may be seen by designers more as a hindrance than a help.

5. DFM work is sometimes performed too late in the design cycle. It must become an integral part of the design process.

6. Considerable work needs to be done on maintaining up-to-date manufacturability rule checking tools in order for Design, Producibility and Process Engineering to be able to find real errors on designs.

Some initiatives can be taken to remedy this situation, for example:

1. Develop tools to capture and apply DFM knowledge.
2. Increase Process Engineering knowledge among designers by providing training on existing DFM tools.
3. Make DFM tools an integral part of the design cycle.
4. Make plant-specific DFM analysis for design checking available and ensure it is used.
5. Enhance existing producibility software tools.

Concurrent product and process design provides the fundamental basis to design products which are producible, reliable and easy to maintain. Products with these characteristics are those with the lowest cost and the highest quality. Electronics design companies work within strictly controlled design processes, variously called gated or phased design processes (Figure 2.11).

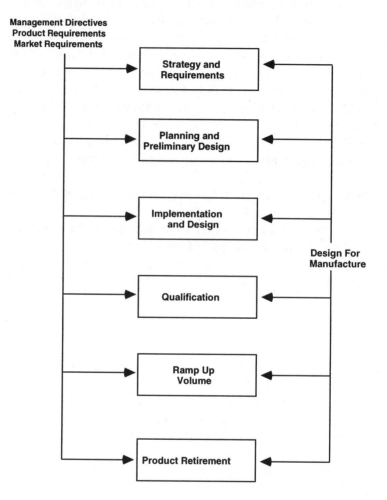

Figure 2.11 *Standard Product Introduction Phases*

Due to the extreme pressure to minimise costs and time-to-market, while maintaining the high quality and reliability expected of modern electronics products, opportunities for application of DFM/A principles must be exploited at every opportunity. This leads to a continuous DFM process taking place throughout the product design cycle (Figure 2.12), which is in essence a tight feedback loop between manufacturing and design. The module layout process (Figure 2.13) represents an area where interaction between product design and production capability is particularly evident and must be understood and optimised.

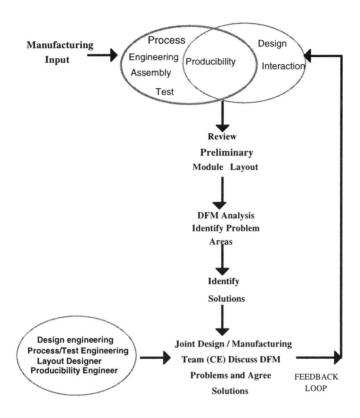

Figure 2.12 *DFM Process Flow*

Depending on the size of the enterprise there may be specialised personnel charged with introducing the product to manufacture, who will work with design and process engineering to identify and resolve DFM issues. Often a DFM log will be maintained to capture and track DFM issues. Resolution of all issues on the DFM log is required prior to sign off of the design by production.

The purpose of the DFM log is to document the agreements made between Design Engineering and manufacturing and test groups which affect the module design. It lists all issues which will affect the manufacture and test of the modules in a volume manufacturing plant. The log is updated and published on a regular basis throughout the design cycle. The log is issued to the Manufacturing Engineering team and Design Engineering.

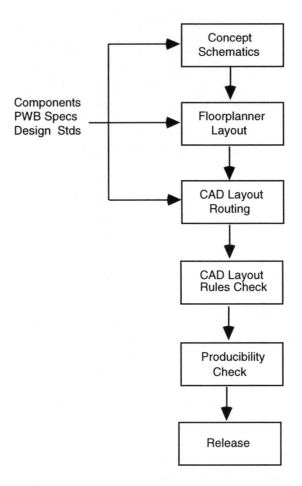

Figure 2.13 *Typical Module Layout/Producibility Process*

Timely resolutions to the issues raised by the team may require that regular CE team meetings be held between the appropriate functions to come to agreement on the issues. The functions involved in most cases will be Design, Producibility, Test, Process, and Quality Engineering.

The DFM engineer has the following responsibilities:

1. Make recommendations on component selection, board technology, manufacturing process yield, and process equipment selection.
2. Work in consultation with concurrent engineering team, on resolution of process issues.
3. Input manufacturing plant requirements to Design Engineering.
4. Review manufacturability analysis reports and feedback concerns to design

PCB Layout Design Systems

Today's PCB layout systems comprise sets of interactive layout tools that give the PCB design engineer great flexibility in making design layout decisions, promoting diversified design processes and fast technical innovation. However, they also permit greater deviation from established manufacturing rules.

As PCB manufacturing processes becomes increasingly automated, requirements for manufacturing become more critical to the design process. Recognition and implementation of manufacturing requirements throughout the design process calls for improved DFM processes and tools. Problems can occur at any stage during the design process. It is both logical and practical to detect these problems as early as possible during the design cycle, so that the designer can take corrective action.

In larger companies, to support the design for manufacture approach to PCB layout, various rule checkers are provided to help the designer determine if the design conforms to manufacturing rules. These apply fixed sets of rules, usually based on corporate standards, to CAD outputs. The exceptions identified by such tools form an important input to the DFM log.

However most small companies do not have access to such tools. Furthermore, the rules contained within them are usually static, and cannot easily be customised for additional manufacturing types. There is also no mechanism for incremental capture of manufacturing rules and decisions. An additional and increasingly important requirement for DFM in electro-mechanical design is the need to be able to combine electronic and mechanical design data for analysis with regard to interactions such as clearance between components and their housing. This can only be achieved in an environment which supports joint modelling and analysis of electronic and mechanical components and assemblies. The architecture described in this book (Chapters 3 to 6) achieves this and other demands in the DFM/A analysis of modern complex electro-mechanical products.

2.4.7   Design for Electronics Assembly

Design for Manufacture in electronics manufacturing focuses mainly on assembly and test. The two main types of assembly currently in common usage are Surface Mount Technology (SMT) and Plated Through Hole (PTH). Boards may be a mixture of both, although many small manufacturing companies are now equipped with their own SMT lines. Table 2.1 lists some of the advantages and disadvantages of SMT relative to PTH.

Table 2.1 *SMT in comparison to PTH*

Advantages	*Disadvantages*
Increased circuit density (reduced product size / increased function)	Difficult to test
Functionally superior (higher speed, lower noise, etc.)	Not all component types available in SMT form
Higher levels of process automation	Higher component cost and less standardisation
Potential for reduced product cost	More difficult to rework
	Soldering process and materials are more sensitive

Due to the spacing issues and high speed automation involved in component placement, most companies involved in SMT manufacture have strict manufacturability criteria in place governing a variety of issues including:

1. Features to enable automatic board / component orientation
2. Preferred orientation of components on-board (e.g. Figure 2.14)

3. Inter-component clearances. These rules are governed by both process imposed limitations (e.g. SMT reflow, Figure 2.15) and equipment limitations (e.g. placement)
4. Component height.

Direction of Processing

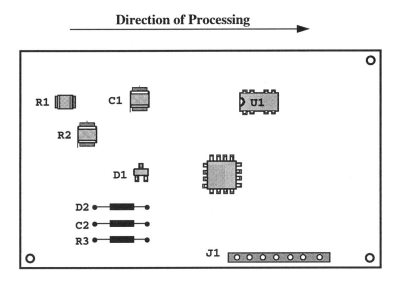

Figure 2.14 *Sample Component Orientation Preferences*

Further examples of PCB DFM rules illustrating typical constraints imposed by assembly and test processes, as well as environmental considerations are given in Table 2.2.

2.4.8 Design for Testability

It is estimated that the cost of testing a typical PWB is about one-third the total manufacturing cost (O'Grady *et al.,* 1992). Therefore any benefits brought about by Design for Test can have a major impact on the whole manufacturing process. Testing of double-sided boards adds further cost, while the testing of SMT boards can be up to 50% more expensive than for pure PTH boards.

The most widely used test method for populated boards is In-Circuit-Test (ICT). This uses specially designed fixtures (incorporating probes as a *bed-of-nails*) to connect to the test points on the board. Through

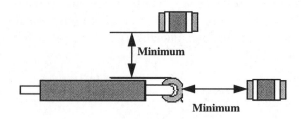

Figure 2.15 *SMT - PTH Axial Lead Clearance for Reflow*

access to the test nodes, diagnostic software is run to apply test conditions to the test nodes and check responses. Accessibility to test nodes on the PWB is a problem that has increased as the density and the complexity of components on the board has increased (Gallagher *et al.*, 1994). This complexity was further enhanced by the advent of fine pitch components, SMT and double-sided boards. The consequent use of smaller probes tends to reduce the quality and reliability of the test further.

Traditionally, the test engineering function starts when the PWB design phase has been completed. This means that test problems, addressed later in the product development cycle, may force modifications to the design, increasing lead times and compromising functionality and quality. Unforeseen problems requiring specialised test fixtures could add up to four weeks to the test fixture production time (Gallagher *et al.*, 1994).

Table 2.2 *PCB DFM Rules*

Rule	Constraint Type
IF lead-pitch <= 0.025" THEN clearance around component package >= 0.15"	Placement
IF board to be wave-soldered THEN all side 2 components must be >= 0.25" from board edge	Wave Solder
IF number of pins > 44 THEN local fiducials diameter 1mm required	Placement
IF a test via is attached to a solder pad THEN distance from edge of footprint pad to edge of via >= 0.25"	Test
IF soldering process to be used THEN solder must be removable using CFC-free solvents	Environmental

Test fixtures are produced mainly by specialist companies, supplied with CAD data by the PWB designer or produced. Problems can arise due to incomplete design data and insufficient knowledge of the manufacturers test capabilities. Consideration of testability problems at the design stage would not only enhance communication, but lead to a higher quality test fixture in a shorter lead time at lower cost. Typical test guidelines (Figure 2.16) cover areas such as:

1. Observe minimum test point spacing limits imposed by bed-of-nails type testing.
2. Components on the top and bottom of the board to be tested must be connected to test points on the bottom of the board.
3. Observe minimum test land areas (imposed by accuracy of test probes position).
4. Observe minimum test land to SMT land spacing imposed by reflow soldering.

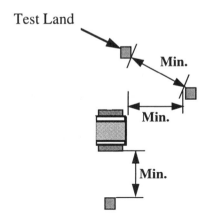

Figure 2.16 *Sample Testability Requirements*

2.4.9 Machining

Design for Machining refers in general to the ease and cost of material removal operations. Examples of such processes used in hole making are:

1. drilling
2. milling
3. grinding
4. honing
5. boring

The type of machining process used, for, say hole making, will depend on a number of variables, such as:

1. Tolerance (see Table 2.3)
2. Material
3. Surface finish required (see Table 2.4)
4. Amount of material to be removed (see Table 2.5)
5. Number of holes to be made

Table 2.3 *Tolerances Associated With Manufacturing Processes*

Process	*hole diameter*	*Tolerance (mm)*	*Tol Grade*
Drilling:	6 - 12mm	0.1	IT11 IT12
	12 - 20mm	0.18	
Reaming:	6 -25mm	0.02	IT7 - IT8
	Over 25mm	0.035 upwards	
Milling:Gang		0.8 - 0.12	IT8 - IT10
Small Slots		0.05 - 0.08	
Turning: Capstan &	to 18mm	0.05	IT8 - IT10
Turret Lathes	25 - 50mm	0.10	
Turning	Over 50mm	0.12 upwards	
Broaching:	Up to 25mm	0.02	IT7 - IT8
	25 - 50mm	0.04	
Honing	up to 50mm	0.01 - 0.016	IT6
Grinding:	up to 25mm	0.007 - 0.012	IT5 - IT6
	25 - 50mm	0.012 - 0.016	
Lapping, machine		0.002 - 0.01	IT4 - IT5
Lapping, standards, reference gauges etc.		Less than 0.002	IT01 - IT3

Table 2.4 *Tolerances & Surface Finishes for Hole Making Processes*

Process	Tolerance (+-)	Finish RMS Values (.001 inch)
Pressure-Die Casting	0.025 - 0.1	500 - 250
Drilling	0.002 - 0.025	250 - 100
Reaming	0.0003 - 0.01	100 - 20
Boring	0.00015 - 0.0025	250 - 20
Grinding	0.00015 - 0.0025	100 - 10
Lapping	0.00015 - 0.0015	20 - 2
Honing	0.00015 - 0.0015	20 - 2

Parts to be machined may have been produced by casting, forging, extrusion, powder metallurgy, etc. The closer the blank to be machined to the final shape desired, the fewer the number and extent of machining processes required. Such net-shape manufacturing is of a major significance in minimising costs. It is also important that the sequence of operations to be performed be minimised so as to reduce the need for moving of parts, changing tool set-ups between operations and the need for additional fixtures. Examples of design guidelines to facilitate tooling are shown in Table 2.6.

Table 2.5 *Sample Capabilities of Drilling and Boring Operations*

Tool Type	Diameter Range	Hole Depth/Diameter	
	(mm)	Typical	Maximum
Twist	0.5 - 150	8	50
Spade	25 - 150	30	100
Gun	2 - 50	100	300
Trepanning	40 - 250	10	100
Boring	3 - 1200	5	8

Drilling

There are various types of drilling machines available:

1. **Portable**: used for small holes in small or awkwardly shaped workpieces.

2.   **Drill press**: usually designated by the largest workpiece diameter that can be accommodated on the table (typically form 150mm to 1250mm).

a)   *Sensitive* - smallest of the fixed machines; used for small diameter work.

b)   *Column* - used for larger diameter holes. The size of work that can be accommodated is limited by the distance between the spindle and the column. Can be multi-spindle. Depth of drilling controlled by the spindle settings (can be as large as 3m). Inconvenient for large workpieces, or when a large number of holes are to be drilled because of setup times.

c)   *Radial arm* - used for large workpieces and when a large number of holes have to be drilled. Less rigid than column-type machines. Depth of drilling controlled by the spindle settings.

d)   *Universal* - heads can be swivelled to drill holes at an angle.

e)   *Lathe* - drilling (holes of moderate size). There is a tendency for the drill to wander from the desired centre line; this may be counteracted to some extent by drilling a small pilot hole first. However, for accurate work a boring tool or boring bar is more satisfactory. The speeds, feeds, and depths of cut depend upon the type of work, the workpiece material, the finish required, and the lathe being used. Types of lathe machines include:

i)   bench
ii)   engine
iii)   turret
iv)   automatic screw

Recent developments in drilling include CNC, three-axis machines in which various drilling operations are performed automatically in the desired sequence with the use of a turret (holds several different tools). Machines with multiple spindles are used for high production-rate operations. These machines are capable of drilling as many as 50 holes of varying size, depth, and location in one step, and are also used for reaming and counterboring operations.

Table 2.6 *Design for Tooling Guidelines*

Rule	Constraint Type
Ensure provision of adequate clamping facilities.	Improvement of quality
Design for a preferential sequence of operations that does not necessitate the re-clamping of components.	Reduction of costs, improvement of quality.
Design-in the adequate tool clearances.	Improvement of quality.

Common Types of Drill

1. **Twist** - standard point twist drill is the most common. Chip-breaker features ground along the cutting edges are available on some drills. This is important when drilling with automated machinery where disposal of long chips without operator interference is important. Other type of twist drill include:

    a) *Step* - produces holes of two or more different diameters.

    b) *Core* - used to make an existing hole larger (has a square end).

    c) *Counterboring and counter sinking* - produce depressions on the surface to accommodate the heads of screws and bolts.

    d) *Canter* - used to help start a hole and guide the drill for regular drilling.

    e) *Spade* - used to produce large and deep holes (have removable tip or bit). They have the advantages of higher stiffness, ease of grinding the cutting edge, and lower cost.

    f) *Crankshaft* - used for drilling deep holes (have good centring ability).

2. **Gun** - used for drilling deep holes (originally developed for drilling gun barrels). Hole depth-to-diameter ratios can be 300 or higher. Cutting speeds are usually high and feeds are low.

3. **Trepanning** - used to make disks up to 150mm from flat sheet or plate. It can be done on lathes, drill presses, or other machines using single-point or multi-point tools. Gun-

trepanning uses a cutting tool similar to a gun drill except that the tool has a central hole in it.

Operations and Cutters

A number of different type of drill and drilling operations (Figure 2.17) are commonly used, depending on the end result required. These include the following:

1.   Drilling - using a standard drill bit.
2.   Reaming - used when smooth accurate holes are required. The hole is first drilled slightly below the finished size using a twist drill, and then opened out with a reamer which is passed through at low speed, and often lubricated with oil. There are two types of reamer:
      a) fixed (used for a particular type of hole)
      b) adjustable.
3.   Counterboring - there are two types:
      a) bar and loose cutter (made from an existing drill)
      b) specially made.
4.   Spotfacing (shallow counter bore usually produced with a counterboring cutter)
5.   Countersinking
6.   Tapping

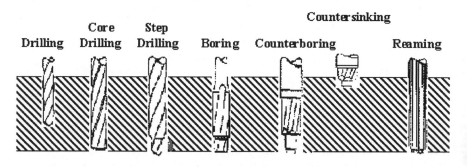

Figure 2.17 *Types of Drilling Operations*

The following are several design guidelines for drilling, reaming and tapping of holes:

1. Components should have flat surfaces, perpendicular to the drill motion; otherwise the drill tends to deflect and the hole will not be located accurately (see Figure 2.18). Exit surfaces for the drill should also be flat (see Figure 2.19).
2. Interrupted hole surfaces should be avoided or minimised for better dimensional accuracy.
3. Hole bottoms should match standard drill point angles. Flat bottoms or odd shapes should be avoided.
4. As in boring operations, through holes are preferred over blind holes.
5. Hole depth should be minimised for good dimensional accuracy.
6. With large diameter holes the workpiece should have a pre-existing hole, preferably made during fabrication of the part during forming or casting.
7. Parts should be designed such that all drilling can be done with a minimum of fixturing or without repositioning of the workpiece.
8. Reaming blind or intersecting holes may be difficult because of possible tool breakage. Extra hole depth should be provided.

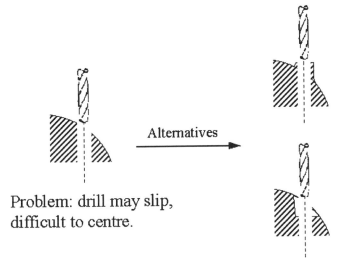

Alternatives

Problem: drill may slip, difficult to centre.

Figure 2.18 *Drilling surfaces should be flat*

Milling

A number of different type of milling machines exist for use today:

1.   Column-and-knee. The spindle to which the cutter is attached may be horizontal for slab milling or vertical for face and end milling, boring, and drilling operations. Most common types are:

   a) Bed
   b) Planer-type - similar to bed. They are equipped with several heads and cutters to mill various surfaces. Used for heavy workpieces and are more efficient than when used for similar work.
   c) Rotary-table - similar to vertical milling machines and are equipped with one or more heads for face-milling operations.
   d) Copy milling - uses tracer fingers to copy parts from a master model. They are used in the automotive and aerospace industries for machining complex parts and dies (die sinking).
   e) Profile milling - has five-axis movement.

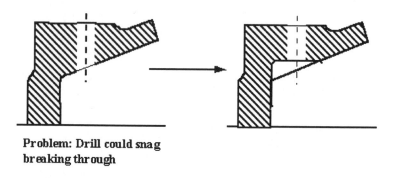

**Problem: Drill could snag
breaking through**

Figure 2.19 *Drill exit surfaces should be flat*

Various milling-machine components are being replaced with CNC machines. These machines are versatile and capable of milling, drilling, boring and tapping with repetitive accuracy.

Grinding

Grinding is used for semi-finishing and finishing. The grinding wheel may be end-cutting or peripheral-cutting. A grinding wheel consists of the abrasive, the bonding material, and the voids. For correct cutting action, it is essential that the wheel wears down during cutting. The correct choice of abrasive, size of abrasive, bonding material, bond strength, and void characteristics is important for correct cutting. The behaviour of the wheel during cutting depends on the area of contact, the work speed, and the wheel speed. The different types of grinding machines include:

1. internal
2. external
3. surface
4. universal

Special grinding wheels are used for gear grinding, thread grinding, and cutter grinding.

Honing

Honing is used to collect local irregularities such as ovality, waviness of axis, or non-parallelism of cylindrical features, and to develop a particular texture. It is used primarily to give holes a fine surface finish. The most common types of honing machines are:

1. internal (these are machines of the vertical type for honing normal length holes, but long holes are honed by horizontal machines.
2. external - using horizontal machines.

Boring

Boring consists of producing internal profiles in hollow workpieces or on a hole made by drilling or another process and is carried out using tools similar to those used in turning. The main types of boring machines include:

1. Vertical spindle - similar to a lathe but with vertical axis of workpiece rotation. Cutting tools usually single point and made of M-2 and M-3 high-speed steel and C-7 and C-8 carbide. The head can be swivelled to produce tapered surfaces.
2. Horizontal spindle - cutting tool is mounted on a spindle which rotates in the headstock capable of both vertical and longitudinal motions. Drills, reamers, taps and milling cutters can also be mounted on the spindle.

Workpiece diameters are generally 1 - 4m but 20m diameters can be machined in some vertical machines. Machine capacities range up to 150kW and can have CNC to improve productivity. Cutting speeds and feeds are similar to those for turning. Design guidelines for boring are generally similar to those for drilling, some examples of additional constraints being:

1. The greater the length-to-bore ratio, the more difficult it is to hold the dimensions because of the deflections of the boring bar from cutting forces.
2. Interrupted internal surfaces should be avoided.

2.5   Currently available manufacturability analysis tools

Although there are a myriad of R&D projects around the world in the area of automated DFM/DFA tools, there are very few commercial solutions available. There are a number of reasons for this, not least the fact that to be really effective, a DFM solution must be customised for the industry and company in which it is applied. As we will see in Chapter 4, knowledge engineering in any domain can be a time-consuming and costly process, and requires specialist personnel. Therefore, while generic solutions, such as the ones we are about to discuss, can improve DFM awareness and practices within a company, production, implementation and maintenance of a customised DFM expert system is beyond the scope of most companies. One of the principal aims of this book is to demonstrate an architectural approach which will overcome many of these problems.

2.5.1 DFMA - Design For Manufacture & Assembly

Formal DFM was created by Professors Peter Dewhurst and Geoffrey Boothroyd in 1983. Based initially on comprehensive work studies, relating part characteristics to handling, fitting times and degrees of difficulty etc., their work was the genesis of the concept of "scoring" designs for DFM / DFA. Now licensed and distributed widely around the world, the Boothroyd and Dewhurst methodology was the inspiration for many of it's successors.

In essence, unstructured DFM/A guidelines and rules are transformed into a structured form that can be applied using paper-based spreadsheet type analysis or computer software. For DFA, for example, Boothroyd Dewhurst uses a formalised step-by-step process to produce a design metric that assesses the design efficiency related to its assemblability. The main steps involved are based on general DFA concepts:

1. Choose assembly method, either manual or automatic
2. Analyse the design for ease of assembly
3. Improve the design and re-analyse

Boothroyd Dewhurst's, Inc. (BDI) DFMA software is designed to be used at the concept design stage. The aim is to optimise design for manufacture and assembly before commitment to detail design and manufacture. The DFMA package contains a Design For Assembly (DFA) module and a Design For Manufacture (DFM) module.

Normally the DFA module is used first, working from a concept drawing. The aim is to improve design efficiency by elimination of unnecessary parts, and choosing more efficient assembly methods. This is assisted by DFA, which provides estimates of times and costs for assembly, as well as other operations such as handling. Through redesign (the DFA module will make some suggestions for design alternatives), the design efficiency figure is improved. This provides a relative measure between design alternatives which should influence the final design concept chosen.

In fact the design team may develop several design concepts with widely differing component parts and manufacturing and assembly operations. The DFM module is used here to evaluate design alternatives from a manufacturing cost perspective, with the design team ultimately deciding on which factors win out in the successful design.

The manufacturing processes which can be analysed by the BDI DFM modules include Sheet Metalworking, Machining, Injection Moulding, Die Casting and Powder Metallurgy.

Using a worksheet the product structure is captured, leading to a possible associated set of assembly and other related operations, such as orientation. For each part in the product structure, the DFA software poses questions on Functionality, Size and Symmetry, Handling Difficulties, and Insertion Difficulties. The answers are compared to a large database of time and motion based data, which assigns tool acquisition time, handling time, insertion time, and then total time for each part. These times are maintained in the worksheet by the software.

Using data input by the user, the software is able to provide analysis in terms of assembly time and cost for the product design being analysed. Suggestions for redesign are made by the software, which the design team is free to incorporate or ignore. Whatever design changes are made however, the same basic sequence of analysis is undertaken to produce the next generation of the design, and comparisons made. The latest DFA Version 8 incorporates the latest windowing technology to provide an improved design structure modelling and analysis environment.

The Boothroyd and Dewhurst Design for Manufacture software consists of five different early cost estimating modules: Machining, Injection Moulding, Sheet Metalworking, Die Casting, Powder Metal Parts. Each module is based on the use of material, process and equipment databases to provide cost information about a particular design. By varying material and/or equipment choices, the user can explore the cost impacts of design alternatives. For example, taking the DFM Early Cost Estimating for Machining, using a library of machining process, the user applies relatively simple parameterisation of parts to estimate part cost. Process variables such as speed and feed rate can be varied, as well as the design variables, including material type to experiment with different combinations. The end result is a process chart of machining sequences, hopefully optimised with respect to the parameters the user deems most important (normally overall component cost).

Other tools produced by Boothroyd and Dewhurst, namely the Design for Serviceability (DFS) software and the Design for Environment (DFE) software simulate the disassembly of products both during and at end-of-life, to help explore serviceability and cost benefits and environmental impacts. As the assembly and eventual disassembly sequences are almost mirror images of each other, it would make sense to perform these analyses concurrently for total life-cycle design.

2.5.2 *Lucas*

Recently Lucas Engineering & Systems Ltd. of the UK have brought out a much enhanced version of their DFA software. The Lucas DFM software brings together modules performing Quality Function Deployment, Concept Convergence, Design for Assembly, Manufacturing Analysis, Failure Mode and Effects Analysis and Design to Cost. They are currently developing other modules including design to service, design for recycling and linking to CAD databases. All modules work from a common database, eliminating the need for data transfer and interfacing between modules.

The Lucas QFD module covers the first two QFD phases, and can be driven by a parts list from the DFA module. The DFA module, similar to Boothroyd and Dewhurst's DFA, supports the development of an assembly sequence flowchart and assembly analysis scoring. The manufacturing analysis module uses part design information to select the most cost-effective material and process choice at the earliest design stage. It plots graphs comparing component costs for a variety of manufacturing processes over a range of annual production volumes. Data generated during the analysis can transfer directly to the DFA module where it is displayed on an assembly diagram and worksheet.

The FMEA module covers both design and process FMEA formats and provides for ordered ranking of severity, occurrence and detection values, and calculates the risk priority number. The design-to-cost module lets target costs be assigned to all areas of a design. For other analysis work, assembly and manufacturing costs can be calculated and compared with the original costs. This module also plots a graph of target costs and actual costs against time to show how the target variance changes over the life of the development project.

2.5.3 Other Commercial Systems

In the Hitachi (or GE-Hitachi) Assemblability Evaluation Method (AEM) for DFA analysis, the two criteria of ease of assembly (as a measure of design quality), and the estimated assembly cost are used to distinguish between designs. The AEM does not distinguish between the method of assembly used during the analysis, where a range of symbols are used to represent assembly operations. Hitachi claimed a deviation of ± 10 % in the estimated assembly cost from actual costs for small mass-produced products, with worsening accuracy as product size increases. Thus this method is very much aimed at a mass-production environment.

2.6 Conclusions - Integrating DFM/A into Different Design Regimes

In developing a generic design environment for Concurrent Engineering the requirements for such a system must be stated. The major requirements are described below:

1. It must be a *negotiation environment* within which the product design can evolve from the requirements stage to detailed design, controlled by the product development team. An evolving product design implies that it must be possible to model the product at all stages of detail (and quantitativity) for all product life-cycle phases and trace the effects of design decisions between product stages (e.g. the effects of a change in a design parameter on satisfaction of customer requirements). To support this concurrence and integration between design tools an integrated product model concept will be used.

2. It should provide support for the basic design activities , which include:

a) searching for objects whose attributes satisfy a given specification fully or partially. This implies the use of libraries of products, parts and features which may be interrogated using search criteria such as functionality, material, size, etc..

b) manipulation of design objects, their attributes and relationships and the product structure. This and the previous point suggest that functional decomposition of the design problem should be supported.

c) the creation and editing of constraints. Within the product model, which is central to the CE design environment, it should be possible to manipulate constraints. In the following section more detail on this requirement and how it may be satisfied will be provided. It will be shown later how different classifications and methods of implementing constraints are needed to handle this problem.

d) checking of constraints of all kinds, whether customer requirement type constraints or DFM/A constraints, must be checked whenever desired in order to assess the effect of design decisions on all life-cycle aspects.

3. The environment must be open to integration with other design tools, such as CAD, QFD, or FMEA. For example it should be possible to create interfaces such that the design environment, QFD system and CAD system are sharing product data. Constraints can be used to capture the QFD matrix relationships and targets, and used to signal violations to the designer.

4. The design environment should provide links with models of manufacturing and other life-cycle processes, so that DFM/A analysis is supported.

5. To support the DFM/A and constraint checking analysis, a knowledge modelling and inferencing capability must be provided which works directly with the product model data.

3. A Generic Systems Architecture

In this chapter we explore the elements which constitute a Design for Manufacture / Assembly systems architecture, and their interaction.

3.1 The scenario system of system design

A systems architecture has three main construction elements: the functional units, the interfaces and the data. Data, whether it is in the form of properties, values or knowledge, must not be embedded. Embedding data produces single purpose, single environment systems of very limited applicability that are almost impossible to maintain.

3.1.1 Manufacturing Aspects

In the current context (manufacturing) the production of items from engineering specifications is concerned with the forming aspects, material conversion and the assembly aspects, i.e. the bringing together of items to form sub-units or complete units. Each of these aspects demands a different viewpoint from the original design viewpoint and the designer must be able to be directed toward or to enter these differing viewpoints during the design process in order to achieve Design for Manufacture / Assembly.

Processes involved in manufacture are categorised in Figure 3.1. Common to all processes is material handling (the outer envelope). Material cutting, forming and casting have been separated out and different processes shown. Assembly has been separated from fabrication - though there are some processes common to each. Treatments and Test and Measurement are the remaining subdivisions. The mix of these processes to achieve a design may vary depending on the facilities available in a particular manufacturing unit. The designer thus must be aware of what may be achieved in house and the economics of out-of-plant operations. Material selected by the designer also has a bearing in the efficiency of resultant processes.

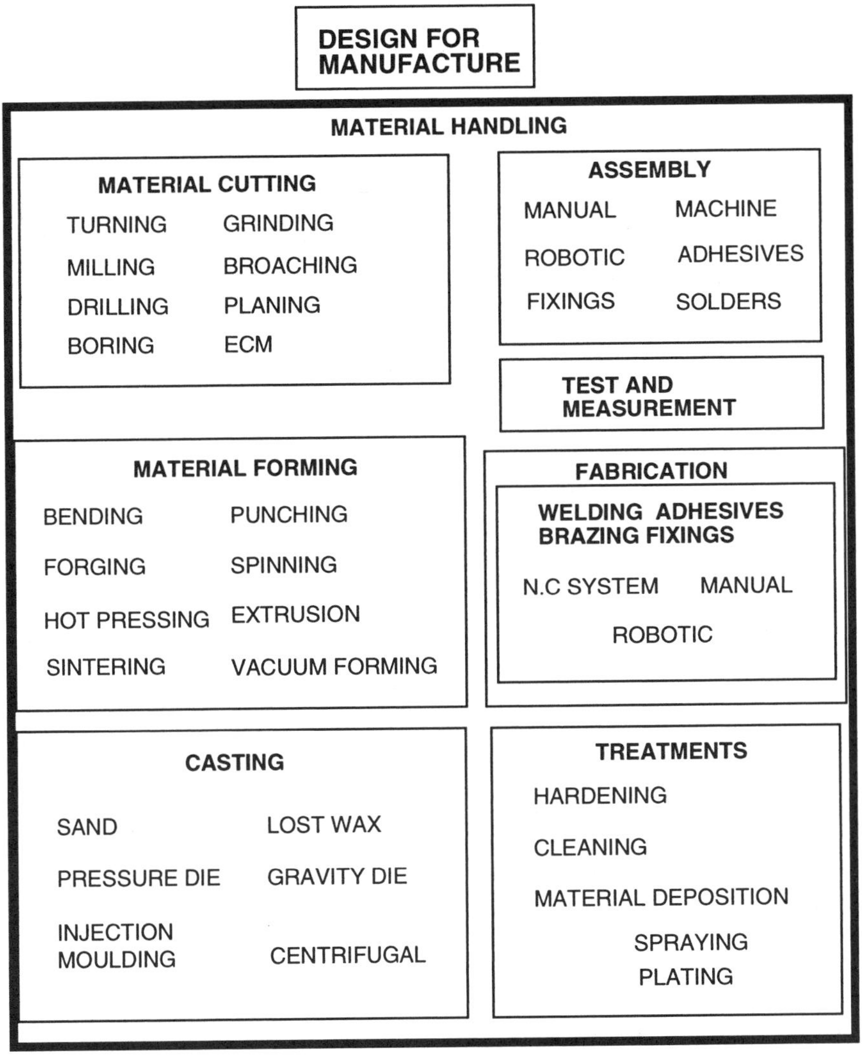

```
┌─────────────────────┐
│ DESIGN FOR          │
│ MANUFACTURE         │
└─────────────────────┘
```

MATERIAL HANDLING

MATERIAL CUTTING

TURNING	GRINDING
MILLING	BROACHING
DRILLING	PLANING
BORING	ECM

ASSEMBLY

MANUAL	MACHINE
ROBOTIC	ADHESIVES
FIXINGS	SOLDERS

TEST AND MEASUREMENT

MATERIAL FORMING

BENDING	PUNCHING
FORGING	SPINNING
HOT PRESSING	EXTRUSION
SINTERING	VACUUM FORMING

FABRICATION

WELDING ADHESIVES BRAZING FIXINGS

N.C SYSTEM   MANUAL

ROBOTIC

CASTING

SAND	LOST WAX
PRESSURE DIE	GRAVITY DIE
INJECTION MOULDING	CENTRIFUGAL

TREATMENTS

HARDENING

CLEANING

MATERIAL DEPOSITION

SPRAYING

PLATING

Figure 3.1 *Processes Involved in Manufacture*

According to British Aerospace, 85 percent of product development cost spend is generated in the first 5 percent of the design cycle. (Philippi, 1991). This type of statement reinforces the argument that design must not be isolated from the subsequent downstream activities needed to achieve a product. Prior to examining the manufacturing aspects in detail, the design process will be analysed.

3.1.2   Design - an analysis

There are many perceptions of design and design for manufacture is just one. Some see design as a purely aesthetic process, others as mechanistic. All these perceptions are recent derivations that would have no doubt amused Leonardo da Vinci. Even though much of engineering design is redressing existing products there is still at its core creativity.

All design may be resolved into three fundamental elements. These are designated aesthetics, strenuosity and kinaesthetics. The aesthetic aspects are these elements of a design that impact the human senses, such as colour, texture, smell, noise and form. Form is also dictated by strenuosity and kinaesthetics. The strenuosity aspects are the load bearing functions of a design such as stress, displacement, current capacity, thermal endurance, wear etc.. The kinaesthetics aspects are the mechanisms of the design - the levers, gears, electrical circuits, hydraulic systems etc., that enable the kinematic realisation of a design. The mix of these elements varies dependant upon the design regime but are present in all design.

Acting upon these core elements of design are at least four processes. Because of the dominance of these processes in contemporary engineering design and the use of computer aided design systems, these processes are often confused as design. These processes are designated reduction, simulation, optimisation and modularisation.

Reduction is the process whereby a design is reduced in "complexity" but still achieves its purpose. Simulation is developing and operating a mathematical model of the design in order to determine how it will function before it is physically realised. A single design may be represented by a number of different simulations. Optimisation is processing a design to optimise selected aspects for example - construction costs, thermal efficiency, material usage etc.. Modularisation is the development of a design from modules. The modules may have been derived from the decomposition of previous designs or designed to suit a family of opportunities. A model of design showing these aspects is presented in Figure 3.2.

In addition a designer/design system must be able to deal with knowledge representation, problem solving and apply perception methods. (A human is very good at situation recognition, that is the ability to take a situation from one context and apply it in a new context). The constraints imposed upon a designer are extremely diverse and are situation and technology dependant. Would an aircraft designer in the 1940's ever have considered building an aircraft from titanium?

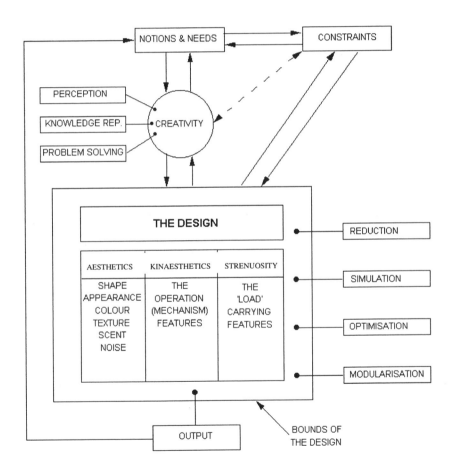

Figure 3.2 *Aspects of Design*

In the design of a generic Design for Manufacturing and Assembly systems architecture particular constraints come to the fore and have a relationship with the four processes. Reduction and modularisation impact upon assembly aspects and the manufacturing plant capability, material properties, production volume all represent constraints. All such aspects must thus be accommodated in an architecture for design for manufacturing and assembly.

3.1.3   Methods of data representation

When a designer and production planning engineer look at a traditional engineering drawing they are each able to interpret the drawing by features, one perhaps by functionality features, the other by manufacturability features. If one however looks at the data structure that represents the drawing or object representations within a conventional CAD system, the data consists of lines, arcs, circles and other basic graphical entities. Their order of appearance in the data structure is related to how the drawing was produced. Designers think in terms of features and objects, not in graphical entities. Links must be established that enable the transparent conversion of the designers concepts to internal computer representations. Thus in order to resolve the problem of data representation some aspects of design theory will be examined.

Design theory provides many axioms (Yoshikawa, 1981, Yoshikawa and Warman, 1987) that may be used to develop an analytical approach towards design. Three axioms are relevant to the arguments developed in deriving an architecture for design for manufacture. The first axiom is the axiom of recognition: this states that any entity can be recognised or described by attributes and/or abstract concepts. The second axiom is the axiom of correspondence and states that the entity set "S" and the set of entity concept (ideal) "S" have a one to one correspondence. The third axiom states that the set of abstract concept is a topology of the set of entity concept.

The first axiom guarantees the observability of entities. It also yields a problem in that the descriptive method for an entity concept must be extensional. The second axiom allows for the fact that "S" may even include entities that will exist in the future. It forces us to check the feasibility or compatibility of the knowledge with the realities. The third axiom signifies that it is possible to operate on abstract concepts logically as if they are just ordinary mathematical sets. General Design Theory aims at three points:

I.   Clarifying the human ability to design in a scientific way;
II. Producing useful practical knowledge about design methodology;
III. Framing design knowledge in a certain formality suitable for its implementation on a computer.

From the human and knowledge analysis viewpoints the extensional data representations arising from the first axiom may be shown to have more advantages than the intentional approach. Current approaches to representing design objects internally within computer systems are intensional. With intensional representation the dependencies between the data become so strong in real objects that it becomes very difficult to change or modify the data schema, whereas in the extensional system data is able to be described by a set of facts (i.e. predicate logic formulae) that are independent of one another. In design typical extensional descriptive forms would be, 5mm square slot, 30 degree chamfer, etc.. Extensional data forms thus may be used so that they equate with design features or even higher order objects such as bushes or gears.

In CAD systems, practicalities require both approaches to be combined. To provide an illustration of the extensional/intensional argument if we wish to chamfer a corner (A), the intensional description causes a change to the data structure that makes it very difficult to judge where the modified data structure is different[1].

The extensional approach would just be to add the predicate - chamfered (A). The effective difference between the two approaches becomes more apparent if one is required to change all chamfered corners to radius corners. Scanning the structure for chamfers having specific properties is a straightforward task. Naturally geometric factors need to be linked to predicates and thus the two structures must have

---

[1] The majority of CAD systems use the data structures derived by Pat Hanratty for his work on computer graphics during the 1960's . The structure consists of lines, arcs, circles and points described by co-ordinate sets and attributes such as thickness, style, colour, etc.. The order of the structure relates to the order in which the drawing was created. Thus a chamfer would be represented as a line with co-ordinates of x^1,y^1,z^1 - x^2,y^2,z^2 and would not have any pointers to associated lines other than common end co-ordinates.

complete computability. (There are in fact several levels of linked structure; consider the screen display and the tracking back needed to satisfy a light pen hit).

This then leads to a further conclusion in that a design system must be equipped with data schema flexible enough to be able to respond quickly to commands. The foregoing thus leads to a concept of duality in that features and predicates are mutually interchangeable. To further extend the usefulness of this concept let object-oriented methods be examined.

3.1.4 The object-oriented approach

The methodology of object-oriented programming is based upon modularity in which functionality and implementation are separated. Prior to object-oriented programming, computer programmes were large sections of code that attempted to achieve completeness. The addition of new elements was non-trivial and was performed by specialists (compare with extensional/intentional data structures). It is difficult to allow for growth in such monolithic systems, user interfaces are inflexible and system utilisation could become ineffective.

In object-oriented systems each component of the system is represented by an object. An object is capable of reacting to messages. A message sent to an object regards the object as its receiver. The message names the type of reaction desired (with its selector) and it may also indicate some other relevant objects (with its arguments). A message returns an object as its value. It is the messages that separate functionality from implementation. Thus an object may have a set of variables and a set of procedures that may be related to a class of objects. The messages themselves may be generated by procedures or from actions occurring in the external world. An object in the final analysis is a package of code but the user may envision it as an icon on a display, a menu item, a section of an engineering drawing, a design feature or a predicate.

The state has thus been arrived at whereby the predicate and their associated features (that are in effect synonymous) can be treated as objects within an object-oriented system. Thus it may be concluded that an object-oriented computer aided design system satisfies the General Theory of Design and is best suited to meet the needs of the designer who thinks in terms of features and their underlying properties. These underlying properties may be treated as objects and could relate to manufacturing or assembly aspects. Thus object-oriented design systems must have considerable advantages in Design for Manufacture applications over graphical-based data structure design systems.

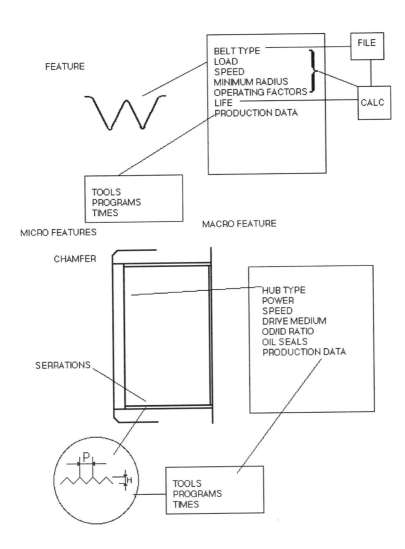

Figure 3.3 *Adding Code to Features*

In many areas of design, accepted common practices (plant or industry specific), safety legislation (national, European or classification agency), and standards (international, national or company specific) are some of the factors that influence the use of standard features. Examples are 25 pin D connectors, automotive clutch housing faces, standard pressure vessel flanges, keyways and splined shafts. Standard features may be incorporated into components and some components may also be standardised elements. Such elements may be considered as high level design features or modules.

The object-oriented approach permits "code" to be added to features. This "code" may for example relate to design data, test results and through to production data as illustrated in Figure 3.3. There is thus the implication that the objected-oriented approach is allowing the design to interact with analysis/simulation programmes and databases by virtue of the designer being able to interact with features.

The object-oriented approach also allows the designer to add modules and their associated links.

3.1.5   Databases

The consideration of design and data representation leads naturally to a consideration of databases. It is well to always consider the plurality of the term databases for there will be databases to suit different needs and purposes within any design hierarchy.

Design data has many-to-many relationships. Without presenting a detailed analysis, this suggests that networked databases are more suited to handle design data than other database forms. Networked databases can handle object-oriented data without any difficulties.

Dependant upon the particular state of the design process, the designer has two kinds of information need, this results in differing query types, expected response, user behaviour and system properties. These features are illustrated in Figure 3.4 and require more attention to be given to the user interface than is normally available for database systems.

The data must also be able to be plugged into the design system shell. One is not only considering "pure" data but also associated knowledge that is in a form acceptable to an inference mechanism.

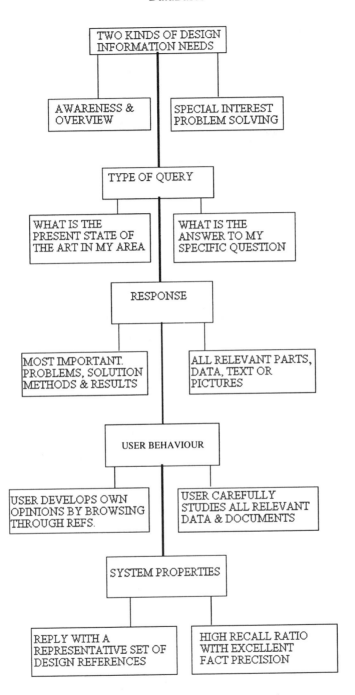

Figure 3.4 *Design Information Needs*

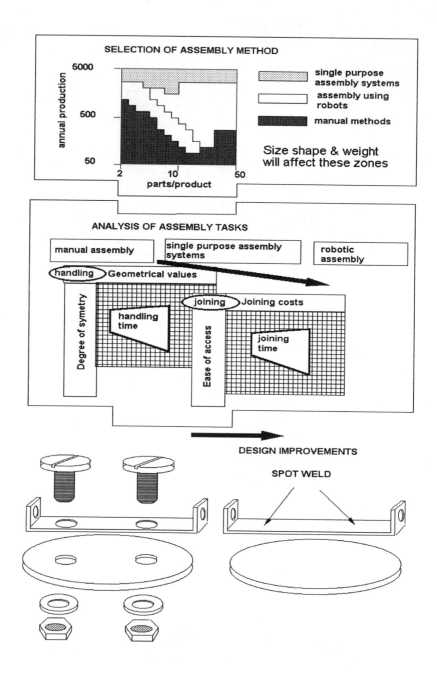

Figure 3.5 *An Approach to DFA Analysis*

3.1.6 Design for Manufacturing

The design system must have a model of the capabilities of the manufacturing system(s) in which the designed products will be realised. As with assembly (Figure 3.5) the production volume and physical characteristics of the product will have a direct influence on the design. Consider for example a simple brass bush; in small quantities it would be machined from bar, medium production volumes suggest a move to near to finished shape brass pressings, while high volumes would benefit from a finished shape sintered bush. A substantial knowledge base is required to support the manufacturing model. This knowledge can be categorised in general manufacturing knowledge and plant-specific knowledge. Both knowledge areas can be further classified into the areas illustrated in Figure 3.1.

As well as the designer initiating a dialogue with the system to arrive at specific results, the system should also be able to initiate a dialogue with the designer. For example the insertion of tight tolerances or precise surface finishing requirements could prompt a dialogue on grinding versus hard turning versus lapping versus honing.

A material Knowledge and Database is another requirement for design for manufacture. Designers must be made aware of machinability factors as well as other material properties. The use of object-oriented methods will enable product structures that will contain total product information. A vital requirement when a design is subsequently modified.

3.1.7 Design for Assembly

All approaches to design of design for manufacture / assembly systems have a product model as one of its elements. This model shall now be considered from the design for assembly viewpoint. The aims of design for assembly are presented in Figure 3.6.

On the one hand the goal is to reduce the number of parts, whilst maintaining functionality, possibly by integration techniques and different manufacturing processes. Limits and fits need to be appraised and the use of facilities that aid assembly (or the deletion of features that hinder). These actions may be seen to relate directly to the previously derived model of design.

On the other hand are activities that are concerned with the development of a product structure that reflects the assembly processes rather than product functionality. (It becomes evident that several different viewpoints of the product model may need to be derived to suit the various aspects of product related processes). Joining activities need to be simplified taking into account subsequent disassembly requirements. At all times standardisation of parts and processes need to be pursued. Influencing these actions is the expected production volume, the number of parts per product and the manufacturing environment. These processes and their relationships are illustrated in Figure 3.5.

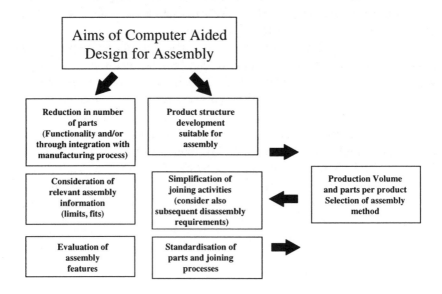

Figure 3.6 *Aims of Design for Assembly*

Production volume, geometric factors parts per product and weight all have an influence upon the selection of assembly regimes. These regimes are perceived as manual assembly, special purpose assembly machines and robotics. The designers attention must - for example - be drawn to the effect that shape changes can have on manual assembly, the effect on robotic gripper design or the ease of orientation in a vibrating feeder. The design for assembly system must provide the designer with information and knowledge concerning the processes and manufacturing plant, with perhaps the ability to simulate the assembly activities.

3.2 The Conceptual Architecture

Prior to considering more "internal" detail, the overall architecture model is presented. In the various information sources that are "plug in" modules, knowledge and data are shown as separate elements. Data is for factual information and knowledge is the meaning and application of the factual information. The diagram in Figure 3.7 shows the conceptual generic architecture derived from the arguments so far presented and is described as follows.

By *design regime* it is meant a particular area of industry or technology as exemplified by machine tool design, automotive design etc. *Features* is the library of features relevant to the particular design regime and encompasses both local and standards elements.

The *product model* is the total and complete representation of the product, this means more than geometrical data, for performance, cost, life and other factors are incorporated.

The architecture is also zoned (refer to Figure 3.7) thus for the design regime knowledge and data and the product-specific knowledge and data are key factors. The action-specific knowledge and data area addresses the three sectors of material , processes and manufacturing resources. It is in this area that some of the previously presented information related to processes will fit, based for example on the classification structure of Figure 3.1. The information in this area will naturally be hierarchical and meet the two types of information needs shown in Figure 3.4. A possible internal structure is illustrated in Figure 3.8 where joining and connecting are considered.

Obviously behind each of the headings is an information structure of considerable depth. The functioning of the architecture depends upon two key elements, the system control and the analysis and integrating engine. Each module will naturally possess their own internal controls but will need to be co-ordinated by the overall system control. The analysis and integrating engine will perform the three functions of adding and modifying the data and knowledge, integrating the information sets and applying inference mechanisms.

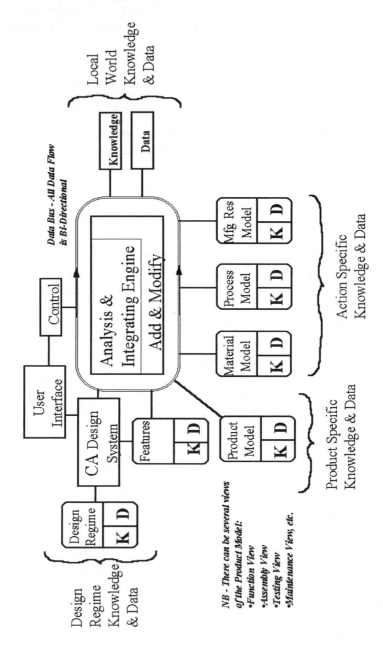

Figure 3.7 *Conceptual Generic Architecture*

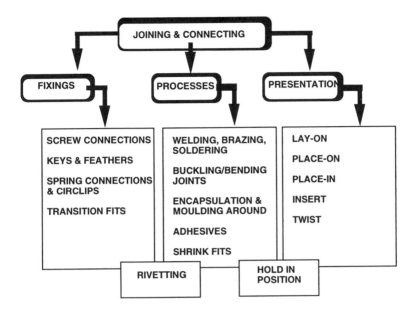

Figure 3.8 *A Sample Process Hierarchy*

3.2.1 *Analysis and Integration and Inference*

As the architecture design work progressed it become apparent that more than a simple inference engine was required. Much knowledge and data in a design for manufacture environment is continually changing and accumulated experience is pushing back the frontiers of engineering performance in an ongoing process. The work on analysing the design process that lead to the conclusions that object-oriented systems are best suited to the aims of DFM/A further suggests that the objected oriented methodology must be applied to the analyses, integration and inference module.

This lead to an examination of DICE (3) and the proposed internal architecture for this module is based upon DICE experience. The structure is illustrated in Figure 3.9. The Meta knowledge (refer to lower part of Figure 3.9) is knowledge concerning knowledge. It directs the system on how to use the knowledge contained within itself. The model is a scheme for dealing with the issue of what pieces of knowledge should be applied. Thus second level trade -off solutions may be handled (the engineering compromise).

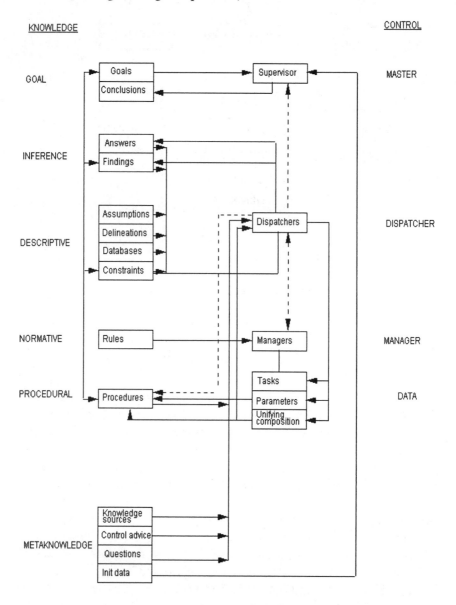

Figure 3.9 *DICE Architecture*

The dispatchers form a hierarchical structure according to the relationships that exist between tasks. This ensures commutative operations, i.e. a change in module (x) is reflected in its mating module (y). If a dispatcher cannot solve a task, the task is decomposed into sub-tasks. The internal architecture is composed of four fundamental components:- (a) the knowledge base, (b) the supervisor that deals with goal attainment and solution network activity, (c) the network that deals with local task solving and (d) dedicated working memory, the whole system being driven by designer consultation.

3.2.2 Interfaces

The architecture for design for manufacture and assembly is held together by its interfaces. The user interface concepts underpinning this work are dealt with in Appendix A. The other interfaces (the connecting lines between the modules) are the subject of continuing standards development activities. For the development of industrial prototypes based on this architecture, the STEP-EXPRESS work was used for design data (geometric) representation and exchange.

The main problem at present is that system developments are outpacing the standards development work. SQL is the current standard in database interaction programming, but will have to be developed further to deal with engineering requirements and the move toward object-oriented approaches. If the design system is eventually to be interfaced to shop floor systems, NC and robotics data transfer interfaces must also be incorporated.

The development of the DEFMAT architecture has shown most clearly the value of object-oriented methods. These methods have value to both the overall aspects and the internal operation and construction of the specific modules. The architecture allows for the coupling together of proprietary systems and elements or the inclusion of specifically developed options.

The study leading to this architecture has indicated that the product structure should be able to be viewed from different aspects. In design for assembly it should be possible to derive an assembly-oriented product structure from a function-oriented product structure as illustrated in Figure 3.10.

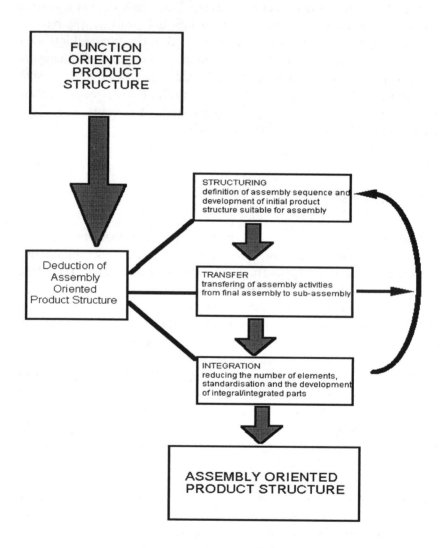

Figure 3.10 *Deriving an Assembly Oriented Product Structure*

3.3 Analysis Engine Concepts

Using an object-oriented and expert system approach, relation and constraint information can be captured as combinations of objects and rules. Thus different views of the same product model can be generated by selectively displaying the necessary information, while the actual data structure itself closely reflects the actual product. Rules can be applied to the product model to check all constraints, or constraints of a particular type (e.g. requirement satisfaction, or engineering conflict) to check whether they are violated. The designer may decide to trigger these rules or they may be triggered by specific design actions, such as the modification of a component parameter.

Other general constraints, which do not vary from product to product, such as environmental constraints, can be represented as rules. These rules in turn will reference objects and their parameters. For example, DFA rules may contain references to specific assembly equipment, modelled as objects. Such rules are linked to the product model by the inference engine of the design environment, which searches the product model for objects whose attributes violate those rules. For example a rule specifying the maximum allowable height of a component through a conveyor, will check the actual component height. The critical height will vary between different machines. Therefore the rule will not hard-code the critical height, but refer to a parameter of the equipment object. Should the equipment be changed, the knowledge can be partitioned such that it refers to the equipment to be used in producing the product.

Thus the product model is vital to support the above types of analyses. It must also support the evaluation of constraints imposed by manufacturing and other requirements (DFX). DFM constraints are usually expressed as rules, and in fact is the most natural way for most people to conceptualise them, i.e. as an IF..THEN type situation. A typical rule in Surface Mount assembly of components might be :

> **IF**   component_side = 2
> **AND**   component_clearance < conveyor_width
> **THEN** increase component_clearance

This rule is instructive, as it shows that the rule effectively breaks down into one or more conditions involving *product data* (component_side), *derived product data* (component_clearance) and *equipment data* (conveyor_width). The conveyor width cannot be known until it is checked that the product will actually pass on that particular conveyor, which implies that *process information* or some primitive process planning is also required in order to associate the equipment model with the product model and some process modelling (which in this case can be simply a list of the processes through which the product will pass). It may also be that there are two process lines, identical except for conveyors of different width, one of which must be chosen for the product assembly.

Inferencing on a DFM rule set can start in two ways. Either a particular hypothesis is suggested, and the inference engine checks whether it is true (backward chaining), or a design change may trigger a chain reaction leading to particular hypotheses or conclusions (forward chaining). In both cases, customisation of the DFM analysis may be effected by exchanging rule sets (or knowledge bases in an expert system implementation).

Thus, as well as segmenting the DFM knowledge into different domains, such as assembly and machining, it is also necessary to provide knowledge segmentation at lower levels, such as the process and equipment levels. Thus the design system must provide the designer with the ability to decide the parameters of the DFM analysis at all levels of granularity, from the highest level, such as deciding to analyse the product against all manufacturing and assembly concerns, down to choosing the particular subsets of the domain which are to be checked.

3.4   The process model

In developing the process model it must be ensured that the resulting system may be configured to deal with a wide specification of processes from a limited specific range through to a complete manufacturing system.

At various stages of the design process the designers may be concerned with specific components and the related processes required to realise them or an entire overview of processes when scheming out new concepts. Thus during the design of a simple component, processes such as turning or drilling may be the only factors under consideration but when considering the assembly of a product of which the component is one element, an entire range of processes would need to be consulted. The process model must therefore be hierarchical. It is also considered to require at least three modes of operation. These modes are envisaged as continuous nudging of the designer, parsing of a sub-component of a design and parsing a complete design. These modes must be able to be invoked by the designer commensurate with his needs. The process model must also provide suitable interfaces to subsequent process planning systems.

The possible hierarchies for the process model levels are illustrated in Figure 3.11. The top level is the plant model the represents the entire plant operation including, if relevant, external suppliers. The next level may be classified as the cell level or section level. At this level the model would contain all capabilities including software aspects.

The third level is the simulation of specific processes. Strong ties exist with CAD systems for simulating cutting processes. Manual machining is still approached using method time measurement (MTM) techniques that have now been transferred to computer systems

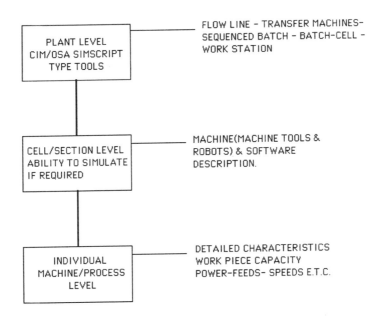

Figure 3.11 *Process Model Hierarchies*

The relationships between specific processes are illustrated in Figure 3.12. These relationships presented in Figure 3.12 are in no way complete but nevertheless will be reflected in rule hierarchies developed.

Design actions are both top down and bottom up. The first activity is the production of an overall scheme for particular concept (top down) that is then realised through the production of a detailed design (bottom up). The process is always one of interaction and refinement. The bottom up approach deals with individual component, groups and components (assemblies) and the total design. The process model should eventually be able to act with all design modes of operation. The more detailed the design stage the more detailed will be the questions and rules.

Often a design is complete before the overall "can I make it" question is asked. Other times the control imposed by manufacturing constraints is such that "new" designs are strongly controlled variations of existing designs. The first aspect is the parsing of a final design whilst the latter represents a limited but highly detailed domain.

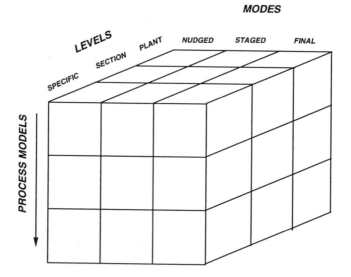

Figure 3.12 *Process Model Complexity Matrix*

The design regime - mechanical - electrical - electronic also imposes specific process constraints (if one refers back to the original model of the design process, processes must represent constraints). The process model must thus consist also of modes and sub-modes to deal with the structured nature of design.

Provision must thus be made for the assignment of priorities (that may be overridden) to deal with the conflicts that may occur between modes and sub-modes or sub-mode and sub-mode. Such an approach will enable a structure to be developed that allows rules to be added without reconfiguration of the entire rule base.

Because the overall system is to be based upon the object-oriented approach certain rules (and other data) may be object-related to other features. Figure 3.13 shows this concept. A vee belt groove - for example - has its related cutting parameters and form tool attached to it, or serrated shafts, key ways and other such features have the manufacturing processes related to them. As new features are developed they and their related process data will be added to the system libraries.

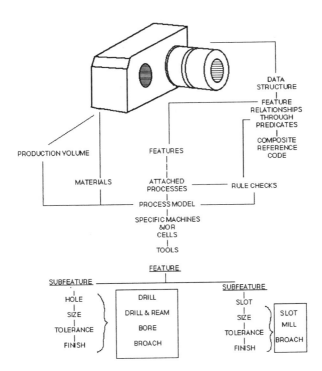

Figure 3.13 *Relating Rules to Features*

Other factors must also be employed to link design elements to the process model. Tolerances dictate processes as illustrated (Figure 3.14 and Figure 3.15) on the data relating tolerances and surface finish to specific processes. Tolerances and geometric tolerances are considered best treated as objects associated with geometric elements. Changing tolerances will not require the model to be reformulated and will only affect those modules that use tolerances as data. Certain parts of the system should operate in a manner analogous to a spread sheet in that data changes invoke a recalculation.

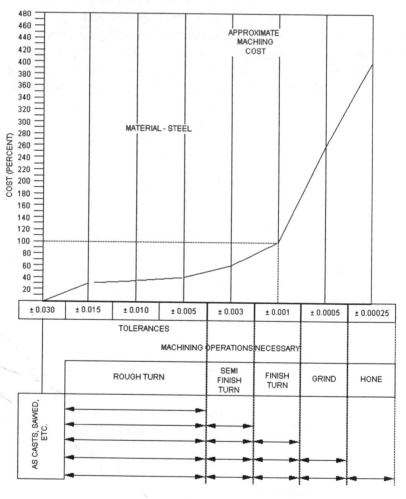

Figure 3.14 *Tolerances and Associated Costs Dictate Processes*

Feature recognition may also have to be part of the process model/design system link for a further viewpoint on design for manufacturing is given by the "traditional" rules. Many feature recognition systems seem to just recognise features that have been input as features, they are really conversion systems. Here we are talking also about features that will result from the inclusion of existing features in a design, i.e. two rectangular bosses may create a convenient "T" form for gang milling or they may create a wrong situation. Provision must be made for the system to ascertain what are undesirable features.

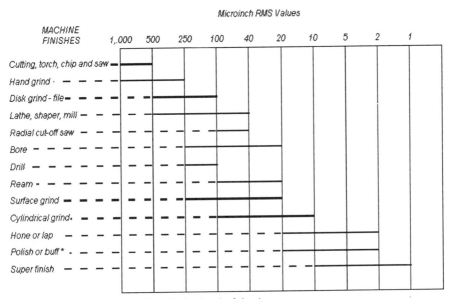

*Dependent on previous surface fish and grit and grade of abrasive

Range of Surface Roughness Produced by Various Machining Operations

Figure 3.15 *Required Surface Finish Dictates Process*

There is thus a wealth of queries that the process model has to deal with. The method by which this information is passed to and from the process model is also a key factor. Another aspect that influences the process selected is the material. The links between the material base and the process model are of importance. The proposals for the analysis and integrating engine which are based upon the object-oriented method are thus in accord with the feature-based approach.

Designing by features and regarding them as objects that are associated with other objects enables the features to carry with them data regarding processes. As the feature library grows so does the power of the system. Rules relating to the use of one feature in conjunction with another will need to be accommodated. A manufacturing feature need not be a design feature but may be created (as previously stated) by the conjunction of two design features. One aspect of designing by features that has been found in the literature is that modelling specialists have a very simplified concept of a design feature and do not thus appreciate how complex many design features are.

Overall geometric form will dictate or influence the process needed to realise that form. Geometric form also dictates the assembly processes needed to build up a set of components.

One method of relating geometric form to the process model is to use Group Technology methodology to derive a code a component. A parsing method to automatically scan the geometry could be employed for this purpose. Sometimes the concept of a design language is given as an approach to this problem. The design language has unfortunately been seen as a user tool. It is suggested that it should be hidden from the user and be an internal representation. Step-Express could be extended for this purpose. If anything can be expressed in sentential form the subsequent parsing, rule checking and error reporting becomes a much simpler task.

Herewith is an example (incomplete) to illustrate this point.

```
Design:=        <Design><Component>I<Component><Component>
Component:=     <Component><Component>I<Features>
Features:=      <Features><Subfeatures>I<Subfeatures>
Sub-feature:=   <Element><Element>
Element:=       <Element>IBossIHoleIShaft...........etc..
Through Hole:=<Circular Hole><Hexagonal Hole><Square
                Hole><OtherShape>
Circular Hole:= Plain I Threaded I Splined I Grooved I Serrated
```

The above arguments reinforce the concept of expressing a design using predicates (which equate to features). There is also a correspondence (though somewhat indirect) with aspects of STEP-Express and a slight extension would allow for the automatic production of total shape codes. Assembly may be approached in a similar manner- *one thought is that if that component codes were sufficiently detailed, assembly could be checked by playing Dominoes with the codes!*

3.5 Control and System Operation

System operation is not seen as a fixed issue but a reflection of the state of use of the system. When the system is newly installed there is a build-up phase, then as the system moves into day to day use there are a variety of possible modes of operation. The philosophy is that the user uses the system to suit the problem regime and manufacturing environment. The user is not constrained by the system. The system is controlled by the user.

The system architecture makes it possible to supply some pre-packaged information to load into the system, e.g. the materials database, some features relative to particular areas of engineering practice, generalised process templates etc.. There is thus a potential service industry for supplying basic elements for DEFMAT type systems.

There are three modes of operation of the "design for" actions. The first mode is a nudging action during the design process. This equates to a spelling checker checking words as they are typed in. The second mode is to invoke checking at the component completion stage and the third mode is the parsing of a complete design. This checking is carried out against a possible three levels of plant model - the total manufacturing process, specific manufacturing sections or cells and the specific machine level.

These modes of activities are conducted under the supervision of the control module and the analysis and integrating engine in conjunction with the user through a powerful user interface. The control module also plays a part in maintaining product model integrity. Locks on the models and databases ensure that unauthorised adjustments are not permitted.

The addition of new features will involve the extraction and linking with the features their related manufacturing data. The specification, for example, of tight tolerances or high surface finishes, will invoke warning messages as to the relative costs of the ensuing operations. The use of object-oriented methods means that features such as tolerances and finishes can be altered without having to reconstruct the basic geometry.

The design regime is the technological base against which the design is performed and it is that collection of knowledge and data, often represented by algorithms, that sets the design values and ensures that safety criteria are met. The design regime must be consulted if any design for manufacturing proposal could violate design operational constraints.

The loading of rules relative to the particular aspects of the system-processes, resources, etc. is also controlled by the control module.

3.5.1   *Control Issues*

The architecture has an inherent robustness in that it is system-independent, modular, extensible, application-independent, able to use templates and has a long life potential. It has also shown that attention must be given, in the future, to the specific design regime, process planning and attaching semantics to features.

In developing concepts for a generic DFM system it became clear that a central control module would be necessary, which would manage the DEFMAT system and its interactions with the user and other systems to which it is linked, such as the CAD system. The overall functions of the control module are to:

1.   manage communications between modules
2.   allow the user to customise the DEFMAT mode of operation, e.g.
        a) alternate between design regimes
        b) choose between different manufacturing environments
        c) decide which events trigger analysis
3.   monitor and react to events in the User Interface or CAD Interface e.g. trigger DFM/A analysis when events specified by

the user occur, such as the creation of a joint could trigger a DFA analysis

4.   allow access, through the User Interface, to database functions such as browsing of data and knowledge, and creation of new design objects.

Through the User Interface, the user gains access either to the DEFMAT control system, or to the CAD system, if it has been opened. The control module, which has its own knowledge and data, uses this knowledge and data to determine its reactions to inputs from the User Interface or other modules.

By allowing the control module to have its own knowledge base, hard-coding of predetermined sequences of action/reaction can be avoided, ensuring a truly flexible control system. This allows the same control system to operate in several different design regimes, such that the interactions between DEFMAT, the user and the CAD system are suited to the design and analysis tasks in hand. The actual analysis sequence is controlled by the Analysis Engine through the knowledge bases, which differ between design regimes.

The customisation of the DEFMAT system is possible by allowing the user to vary the values of System Variables. The values of the System Variables are then used as the basis for decision making during control of the system. The SV's can offer choices to the user, such as choice of Design Regime, or act as switches, such as "Analyse when user creates a new joint / Don't analyse when user creates a new joint". The SV's can also be referenced by the Analysis Engine during analysis, for example to determine the level of output information required by the user. Another example of using SV's is in a PCB assembly analysis, to allow the user to decide which assembly line the product should be analysed against.

The control module also allows the Analysis Engine to access the DEFMAT user interface during analysis, when user input is required, thus providing a homogeneous environment for the user, with all interaction with DEFMAT occurring through the user interface.

4. The Product Model and CAD Interfacing

In this chapter we describe the fundamental modelling techniques used in the definition of the product model required to support the DEFMAT architecture. This leads to a description of the design and manufacturing features used to implement the integration between design (CAD) and the DFM/A expert system (and associated process and manufacturing resource models). There follows a discussion of the CAD interfacing mechanisms required to support a fully generic DFM/A architecture concept.

4.1 Product Model - Structure and Object-Oriented Approach

By "object-oriented" we mean that the software is organised as a collection of discrete objects that incorporate both data structure and behaviour (Rumbaugh *et al.*, 1991). This contrasts with conventional programming methods where data structure and behaviour are more loosely associated. More detailed descriptions of the object-oriented approach are also provided by Booch (1994). There follows explanations of basic object-oriented modelling and design terms and concepts used in the DEFMAT system.

4.1.1 Classes and Objects

An object in this modelling context means an abstraction of an actual real object in the problem domain. This model object has the ability to store information relating to itself as well as interact with other objects. The object has a state, behaviour and identity. Each object is distinguishable from all the other objects of a system. For example

(Figure 4.1) each screw in a product is an object that has a set of attributes and relationships to other objects in the product. A joint object, for example, could require screws. In this case there would be a relationship between screw and joint that is also modelled as an object. Typical attributes of the screw object would be (Parmley, 1977):

1. geometry
2. key engagement
3. flushness tolerance
4. theoretical sharpness

A class describes a set of objects with similar structure and behaviour. In the example in Figure 4.1 both screws belong to the same class "screw".

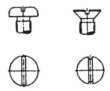

Figure 4.1 *Two Objects of the Class Screw*

4.1.2 *Polymorphism and Inheritance*

Two other important terms in the world of object-oriented modelling and design are polymorphism and inheritance. Polymorphism means that the same operation may behave differently on different classes. Inheritance is a hierarchical relationship among classes wherein one class shares the attributes and/or behaviour defined in one (single inheritance) or more (multiple inheritance) other classes. In our screw example it may be possible to model different types of screws in different sub-classes. In this case the common attributes and behaviour of the all screw classes could be described in one super class "screw" (Figure 4.2).

4.1.3   Modelling Concepts

In the DEFMAT demonstrator prototype the basics of the product model, as shown in Figure 4.3, were designed in EXPRESS-G (Express, 1990), a graphical modelling language that can be used to describe the classes, objects and the relationships among the objects of the problem domain. Figure 4.3 illustrates a part of the product model in EXPRESS notation describing assembly-specific entities and their attributes. The symbols for an entity (class) are solid boxes enclosing the name of the entity. Below we use the term class instead of the term entity. The attributes of a class, for example the class component has the attributes *id*, *descr*, and *name*, are connected by lines to the class they belong to. Lines between class symbols represent the relationships among the classes. In general there are three different relationships:

1.   association
2.   aggregation
3.   inheritance

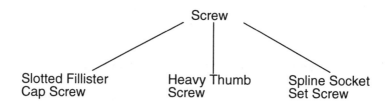

Figure 4.2 *Sample Class Hierarchy*

An association regarding to Booch (1994) is "a relationship denoting a semantic connection between classes". An aggregation can be seen as a special association, such as

1.   Assembly-Parts
2.   Container-Contents

An example of an aggregation is the "assembly-piece_part" relationship. It is modelled as a solid line between the two class symbols. The circle marks the emphasised direction but, generally, these relationships are bi-

directional, except of the inheritance that is displayed as a thick solid
line. In Figure 4.3, the classes "piece_part" and "assembly" inherit all
the attributes and behaviour of the class "component". Additionally,
they do have some special attributes and behaviour. An assembly, for
example, could consist of subassemblies, this is an optional relationship
displayed by a dashed line.

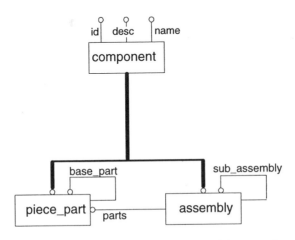

Figure 4.3 *Inheritance, Association and Aggregation*

Most graphical models are large and may span several pages. Therefore
there are symbols to refer between separate pages. In Figure 4.4 such a
symbol (oval box) is shown.

4.1.4 Product Model Structure Overview

The assembly model shown is derived from the analysis of assembly
planning processes of different applications performed during the
DEFMAT project. These processes are directly influenced by the
product, the assembly process, and the resources. Assembly planning
will also depend on several additional influence factors such as order
(quantity and time) and plant layout (or organisation). Therefore the
assembly model contains five base classes:

1. product

2. resource
3. process
4. order
5. organisation

Figure 4.5 illustrates the relationships among these base classes in EXPRESS-G notation.

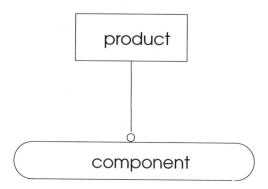

Figure 4.4 *Compositional Symbol*

The work of the designer is influenced by the method of assembly planning used. However, regardless of what method is used, the assembly task is the main thing to be considered. In order to emphasise this, therefore, the joint gets to a central place in the model. A joint is characterised by the parts of a product and their joining relation as well as by tools and the joining method containing both primary and secondary operations (Lotter, 1982).

Operations do have time constraints describing their starting and ending time. Information about this is in the time schedule modelled as a relationship between the classes order and operation. Additionally, operations belong to a specific subclass of the class process (structure). Assembly planning should consider the equipment of the organisation. If you do not have the required assembly tools, you won't be able to connect the parts belonging to the joint. Therefore there is a connection between equipment and organisation.

In general an organisation offers different products on the market that are described in the production program. An order corresponds to one of these products in the product program. In the model the base class product fulfils this conditions by having a relationship to organisation and order. Because of the delivery date of an order there is an connection to operation named time schedule. Additionally, a product consists of different components.

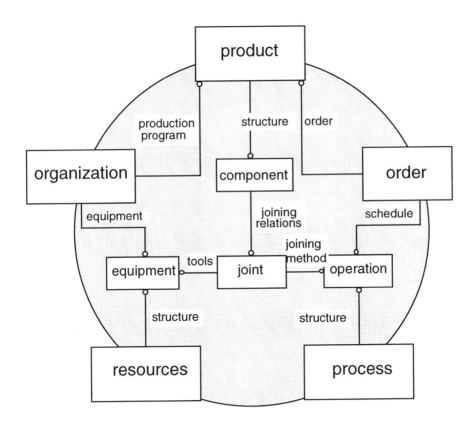

Figure 4.5 *Relationship among the Base Classes of the System*

4.1.5   Detailed Product Model

Figure 4.6 and Figure 4.7 describe the DEFMAT product model in detail. The EXPRESS-G notation is used to model all the relationships between the components of the class product.

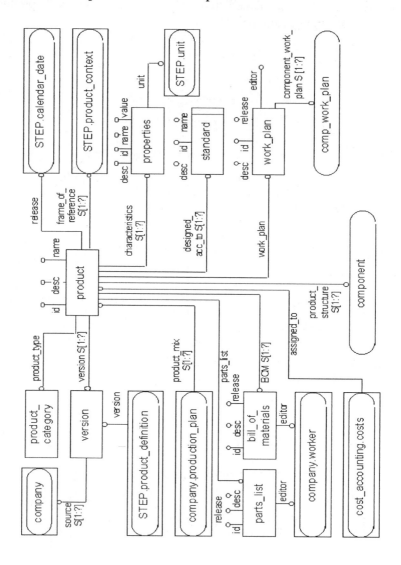

Figure 4.6 *Top Level of the Product Model*

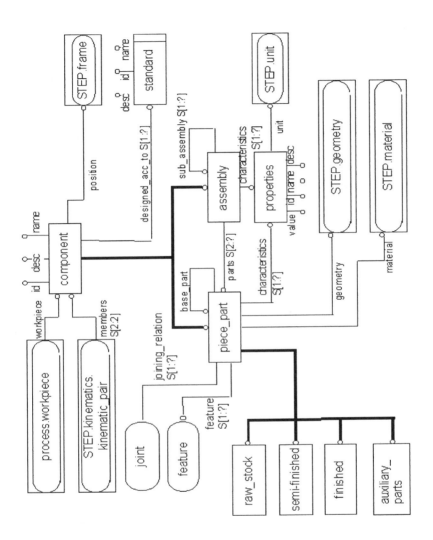

Figure 4.7 *Component Level Model*

4.1.6   *Storage of Object-Oriented Product Models*

A persistent storage facility is the most important aspect of a database but modern databases provide additional functionality, such as management of concurrent users (on a variety of privilege levels and access requirements), and maintenance of data integrity (hopefully regardless of system, media or network failure).

Objects have a life time which is the period of time from the creation to the point where they are destroyed. This time, normally, corresponds with the life time of the process in which the object is created. To make objects persistent means to give them the opportunity to keep their information even if the process in which they are created won't exist any more. This characteristic is achieved by using Object-Oriented Databases (Figure 4.8). Objects that are stored in an Database may be accessed by processes until the objects are explicitly deleted.

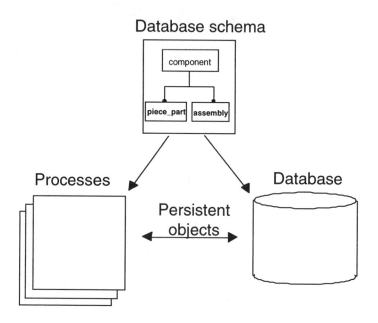

Figure 4.8 *Connection between DataBase and Processes*

In a typical Object-Oriented DataBase persistence is achieved through inheritance. It provides an Object that encapsulates all attributes and behaviour needed to store itself into the DataBase and access information of itself out of the DataBase. Every object in the product model that should be persistent must inherit the attributes and behaviour of this special object called "Object".

The database schema is the class definitions and relationships of the models. The Object-Oriented DataBase works with class definitions specified in the Application Programming Language (e.g. C++) used to implement the interfaces between the system modules. As a logical consequence one of the first steps required in creating the product model - OO database interface is to transform the EXPRESS model into C++ class definitions. A number of commercial translation tools are now becoming available to assist in this process in interpreting STEP Express language to produce C++ code.

In addition to the object methods defined within the STEP Express schema, it is also necessary for the C++ classes to inherit methods which allow their data to be accessed by other C++ modules.

4.1.7   Features in CAD-DFM Integration

The fundamental approach with existing CAD systems is that they are geometrically-oriented. They possess little or no ability to store non-geometrical information which also has a significant influence on design. For example, consider a simple hole drilled in a cube of metal. In a drafting system this would be represented as a circle within a square. This is all the understanding of the part that the computer has. It is only the human designer's interpretation of these symbols, according to certain conventions, which makes them meaningful. The two basic approaches in which features can be used are:

1. Design by features
2. Feature recognition

4.1.8   Feature Representation Methodologies

In the design and manufacturing environment a part can be described in a number of ways, typical methods being engineering drawings (2D), physical models, Group Technology (GT) codes, symbolic representations and the modern computer-based geometric representations: wire frame, surface and solid models. In prototype realisations of this architecture, a parametric feature-based CAD system was chosen as the sample CAD system within DEFMAT as it combines both solid modelling and feature-based design.

4.1.9   Classification of features

A general structure for the definition and implementation of design and manufacturing features, was developed within which several classes of features were defined :

1. Design feature A parameterised geometrical entity used for building the CAD model. Several levels of complexity exist :

a) Functional: complex geometry, with a specific function for a product family. The feature is usually linked with a set of rules, describing the interrelations between the individual parameters. The functionality of a feature can be linked with a single piece part or with an assembly.
b) Compound: complex geometry, with no specific function. The feature is used for rapidly adding geometry without specifying too much detailed information. This type of feature corresponds to the drafting macros used in earlier CAD systems.
c) Basic: the simplest geometrical entities for describing the part. The higher level features can all be built using these basic features. Typically this type of feature is provided standard by a feature-based modeller. The basic Boolean operations on a solid are all performed using this type of feature. As with the more complex types, design rules can be defined, describing the relations between the parameter. The basic design features have a dual role within the DEFMAT architecture. They act both as the building blocks for the more complex elements and as a direct link with the manufacturing features and methods. Therefore rules are linked to each feature to automatically translate the information back and forth the DFMA analysis.

2. Manufacturing feature A parameterised entity linked with one or several alternative manufacturing methods. It is the link between the manufacturing knowledge and the designed product.

3. Assembly feature (or joint) A parameterised entity linked with one or several assembly methods. Typically it consists of one or more design features in the same or several piece parts, belonging to a specific assembly. As with the manufacturing features the parameters of the joint are directly linked with knowledge and data about the product's functionality and the assembly methods.

4.1.10   Hierarchical structure of the features

For an optimum use of this concept, the structure of the features has been split into several areas :

1. Generic knowledge and data for a wide variety of products. This type of feature typically represents the basic or compound features as they are usable for all types of designs. The information linked with the features can be inherited or modified for a specific Design regime or a specific Product.
2. Specific knowledge and data for a product family. Within a product family, specific functions are defined. This functionality is clearly linked with the functional design features. The particular behaviour can be modified by the Design regime.
3. Specific information for a process model

The set of available manufacturing features is defined by the chosen Process model. Therefore the description of the product in terms of manufacturing features will depend on the actual type of process analysed. In the example below two manufacturing views are presented on the same product (Figure 4.9, Figure 4.10). An identical set of design features allows different sets of manufacturing features. The structure is based on definitions made in CAM-I and ISO-STEP feature classifications.

4.2   Interfacing with different CAD systems

Since the product model is the primary source of input for DFM/DFA tools, a close integration with the CAD system enables the analysis of a product while it is being designed. The CAD interface integrates applications with a feature-based solid modeller and allows to extract and manipulate data directly from the CAD product model. As support for different analysis methods (dynamic/static FEA, flow simulation, kinematic analysis, interference control) may be essential, the CAD interface allows integration of tools for various DFM/DFA technologies and calculations with one unambiguous CAD-model..

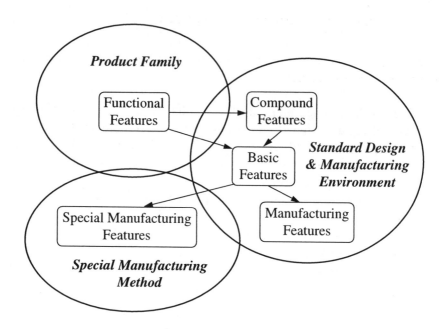

Figure 4.9 *General Structure of the Design and Manufacturing Features*

4.2.1   General functionality of the CAD Interface

The task of the CAD interface in a DFM/DFA application is to translate the geometrical information in a CAD application to the functional information needed for a DFM/DFA analysis. Before one can define the full functionality of the interface, one has to define the objects (information entities) the interface has to work with. Before describing the objects of the CAD interface a description is given of the objects contained in the product model on the CAD system.

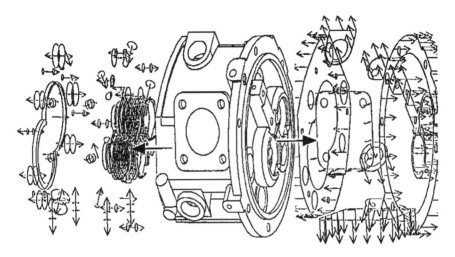

Holes to be drilled

#D 2, hole_through_countersink
   dia, 20.0, -0.01, 0.01
   depth, 15.0
   direction, 0,0,-1.0
   origin, 12.11,1.71,-32.82
   free_bottom_distance 2.31

Faces to be machined

#D 7, hole_through_countersink
   area, 345.01
   minimum_radius, 2.0
   direction, 1.0,0.0,0.0
   surface_roughness 3.2
#T position tol 10,14 44.00, .2,-.2

Figure 4.10 *Manufacturing feature view*

4.2.2 *Interface mechanisms*

Different mechanisms are supported to link an application with the CAD system via the interface. Applications can be linked together with the CAD system or written as external programs. For the latter inter-process communication allows processes to access the CAD DataBase and to extract and manipulate data from the CAD system.

4.2.3   Objects in the CAD system

The objects in the CAD system[2] can be divided into several classes:

Assemblies Assemblies are the most complex entities in the CAD application. An assembly consists of a number of components. A component may be either a part or a sub-assembly, which itself contains a number of components. They are built up by placing together components and contain parts (atomic products), sub-assemblies and references and relations between components (see Figure 4.11).

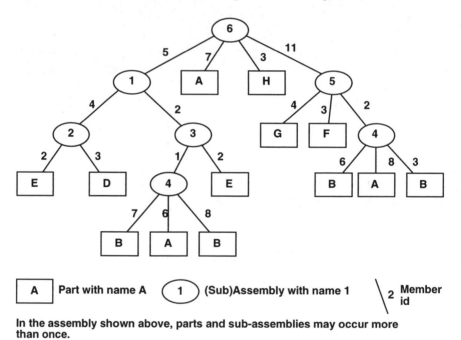

In the assembly shown above, parts and sub-assemblies may occur more than once.

Figure 4.11 *Building CAD Assemblies*

---

2 The objects described here are valid for a parametric solid modeller such as Pro/ENGINEER, as used within the frame work of the DEFMAT project.

The same part or sub-assembly may be placed more than once in the assembly hierarchy. Therefore, for functions that deal with components, it is not enough to identify the corresponding component in the CAD interface by its name only: its position within the hierarchy must also be specified. This can be done by referencing to the arrays of member ids that specify the "path" down the assembly hierarchy to the individual component (see Figure 4.12).

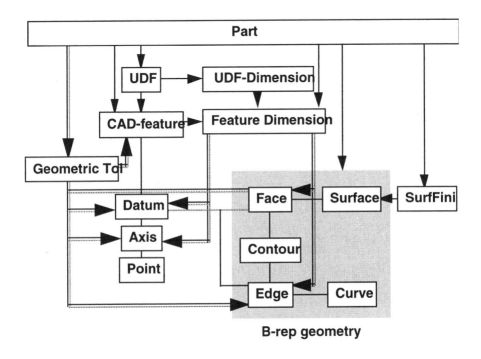

Figure 4.12 *CAD Geometric Hierarchy*

Parts A part is a product that is not manufactured using an assembly operation but using any other type of manufacturing. In the feature-based solid modeller a part may contain CAD system features, that build up the geometry of the a part, User Defined Features, standardised by the user, and relations between features .

CAD system features In a feature based solid-modelling CAD system all geometry is created by features. The built-in functions and features of the CAD system are used to design the part. These features are not standardised by the user and, except for their geometry, they do not contain extra design information. CAD system features contain a feature type, toleranced/non-toleranced geometry and placement dimensions. Additional information can be stored concerning the B-REP and the relations in the part or assembly.

Tolerances The manufacturing of parts and assemblies uses a degree of precision determined by tolerances. CAD tools support several types of tolerances, e.g. dimensional geometrical and surface finishing.

Drawings The basic functionality to document the solid models is provided by drawings, which share two-way associativity with the model. Any changes which are made to the part or assembly will cause the drawing to automatically update and reflect the changes (addition or deletion of features, dimensional changes, etc.). Likewise any changes made to the model in the drawing will be immediately visible in the part or assembly.

4.2.4   Interface mechanisms for applications

Three mechanisms can be used to link a DFM/A application with a CAD system. The DFM/DFA applications can be linked together with the CAD system or written as external programs. For the latter, data can be exchanged in two ways :

a) via a standardised file format: data transfer between the application and the CAD interface is performed via a standardised file format (such as IGES or a proprietary standard).
b) by inter-process communication: inter-process communication will allow processes to access the CAD DataBase and to extract and manipulate data from the CAD system.

Direct integration Application code is linked together with the CAD system's Application Programming Interface (API). Since application and interface are linked together, the application program can make use of all API functions (Figure 4.13).

Data transfer via files External processes, not linked together with the CAD API, can not make use these functions. Data to an external process can be transferred via a standardised file format. If synchronisation between the two processes is required, inter-process communication must be developed to synchronise the transfer of data between the two processes (see next alternative).

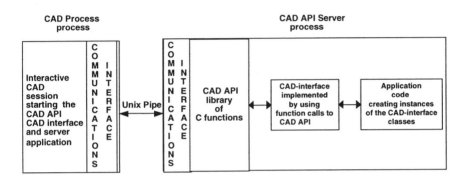

Figure 4.13 *Direct Integration with CAD*

Inter-process communication External processes can use functions of the CAD API by implementing an inter-process communication. In place of executing the code of the CAD interface in the external application process, messages and data are sent to the process running the CAD interface. These messages and data tell the CAD process which functionality to run. After execution the CAD process can return the resulting information to the application process. The data exchange in the interface passes through different stages :

1) CAD implementation This contains the actual API function calls. The CAD interface must be expanded with a module able to receive messages and data and send the results back to the application.

2) Communication interface(s) A communication interface between the application and the process running the CAD interface must be installed. The command and data format for exchange between the two processes must be standardised: on top of all functions a message and data passing system must be defined so that the correct function can be called with the correct data. These interfaces also take care of the synchronisation between the two processes.

3) Application The product structure is identical to the one used with the CAD implementation. Functions that would access the CAD tool actually access the communications layer. In Object-Oriented terms, one can say that the classes have identical public functions but separate private and protected members (Figure 4.14)

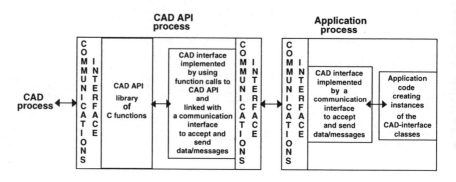

Figure 4.14 *Inter-Process Communication Interfacing*

5. Knowledge Engineering and Inferencing

The design process involves the search for an acceptable solution to a given design problem or specification, within the limitations or constraints imposed by all aspects of the product life-cycle. In this chapter we explore the DEFMAT approach to enabling the application of manufacturing constraints to the design problem. It is important to note that as well as seeking to avoid "over-constraining" the design, we also seek to improve the manufacturing and assembly quality of the design, through offering DFM/A knowledge in a constructive manner.

5.1 Knowledge Elicitation and Acquisition

Most companies operate some form of design for manufacture or assembly procedure, however informal. DFM in practice may take the form of preferred component choices and standard design practices. In fact many medium to large size companies will have formalised DFM/A to some extent through a minimum set of design rules to ensure compatibility with manufacturing process capability. In many cases, however, standard rules are often ignored, as their immediate relevance to a particular design situation may be unclear, or the penalty involved in ignoring the rule may not be self-evident. It is in such situations that expert systems will come into their own. Design checking can be automated, while the impact of violation of DFM/A constraints can be clarified or explained to the designers through simply tracing the inferencing process directly from the design parameters.

In building expert systems it is normal to differentiate between knowledge elicitation and knowledge acquisition. Knowledge elicitation is commonly understood as the process of gaining knowledge from human experts, usually through interview or cased-based techniques.

Knowledge acquisition is commonly understood to be the process of acquiring knowledge from human experts and from other sources such as manuals, books, journals, etc.. Thus knowledge acquisition is the more widely used term, covering a wider area. The role of the knowledge engineer is to perform the acquisition and/or elicitation of knowledge and to represent it in a form useful to expert systems. This task would include fine-tuning roughly formulated knowledge within the expert system to provide the best and most efficient reasoning.

Knowledge acquisition is often seen as the major impediment to producing really useful expert systems. A number of different knowledge acquisition techniques have been document by those engaged in expert systems and knowledge engineering research and development. Ten different techniques are described by Chetupuzha and Badiru (1991), including interviews, protocols, walk-throughs, role play. For an analysis of such methods and their relative merits in different situations, see Hoffman (1987).

5.2   Knowledge Representation

Several methods may be used to represent knowledge and DFM/A rules such as IF...THEN rules, frames, semantic networks, and object-oriented programming (OOP) techniques. In OOP the basic unit of information is the "object" which has a name and a set of attributes. These are quite similar to frames, but have the added concept of inheritance of attributes. Object-oriented programming is widely regarded in the literature surveyed as being a good candidate for the implementation of DFM. In fact Kelly-Sines *et al.* (1989) argue strongly for it's usage, citing the following advantages over procedural languages:

1. Information and relationships are easy to maintain. For example a change made to a parent object is automatically inherited by the child (if the inheritance property is not overridden).
2. It is not necessary to specify the order of execution of relationships in an object-oriented system.
3. The system is more efficient in its use since relationships are executed only if they are needed.
The immediacy and transparency of rule-based representation means it has been the most common approach to date. In modern object-oriented

expert system shells, such as Nexpert™, objects and rules are combined to provide powerful data modelling and knowledge processing capabilities. The knowledge used in an expert system may normally be divided into two types:

1. Declarative knowledge. Normally facts about objects, events and relations will be classified as declarative knowledge. Therefore the product structure and parametric relations between product features can be classified as declarative knowledge.

2. Procedural knowledge. This describes that type of knowledge which, in conjunction with factual knowledge, creates inferences. Thus rules can be classified as procedural knowledge.

Manoharan *et al.* (1990) point out that the examination of human expertise and it's approach to problem solving, i.e. on a more basic level of knowledge acquisition, is the first and most important phase in the development of an expert system. While this is true, it must then be possible to logically represent this knowledge, hence the use of a knowledge classification system. Yoshikawa (1993) also stresses the need for categorisation of knowledge in order to be able to use it.

An example of using an object-oriented design support system for machining is provided by Moriwaki and Nunobiki (1994) using design objects to represent the design process model for design of machine tools. These objects are activated by messages to determine the parameters of the objects to be designed. This is an example of a fairly specific application, but with the interesting aspect of using two types of objects, design **process** objects to perform the design, acting upon design **data** objects which represent the machine tool. However the inherent embedding of manufacturing knowledge within design objects does not necessarily lend such systems readily to use across the design cycle. The representation of knowledge is fundamental to the efficiency of problem solving. Two main forms of representation are:

5.2.1 *Rules*

Three categories of rules and their application are :

1. simple matching   IF a THEN true.
2. two-sided rules   IF a THEN b.
3. relationships   IF a AND b THEN c.

The one-sided rule is effective for a simple statement of facts. The general two-sided production rule can be seen as a cause-effect pair or a trigger-action pair. For knowledge represented in this form, they could be used from either directions. The forward chaining (Figure 5.1) interpretation is good for constraint satisfaction, as in the example of selection of process and equipment. The backward interpretation is for goal searching. If DFM/A knowledge can be represented in this simple form, then the same knowledge regarding process selection could be used to provide design advice. Rules are also very good in explaining the reasoning, explanation is just a simple tracing of the rules fired.

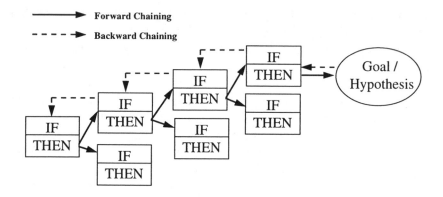

Figure 5.1 *Chaining in Rules-Based Expert Systems*

The complex rule is used to represent relationships and used in particular for goal searching. As a form of organising a large volume of information, rules are not efficient as the structure of information could not be exploited. It is also extremely difficult to maintain when the knowledge-base grows and evolves.

5.2.2   *Objects*

Object-oriented representation is currently the most popular form in different research work. Objects have a uniform structure of attributes, which is very convenient for information maintenance. The hierarchical class structure supports inheritance and data abstraction and hiding. Thus analysis could be conducted at the appropriate level of abstraction. The behaviour of the objects defined in the form of methods greatly enhance the reuse of codes.

The main advantage of object-oriented programming is in software reuse and maintenance. However, the disadvantage is in the difficulty in control of software, as the sequence of programme execution is no longer explicit.

5.2.3 Problem Solving Strategies

Problem solving strategies inherent in DFM/A knowledge *per se* may be categorised as analytic or synthetic. The analytic approach is best suited for diagnostic problems, the synthetic for planning or configuration tasks. In analytic problem solving a solution is found by selection from a finite number of possible solutions. Unfortunately in design of complex objects, the number becomes impossibly large. One analytic approach is heuristic classification proposed by Clancey (1985). It is structured into the phases data abstraction, heuristic mapping onto a hierarchy of pre-enumerated solutions, and refinement within this hierarchy. The knowledge engineering methodology used in DEFMAT follows the procedure developed by Karbach and Linster (1990), as shown in Figure 5.2. This approach has been used in DEFMAT for process selection.

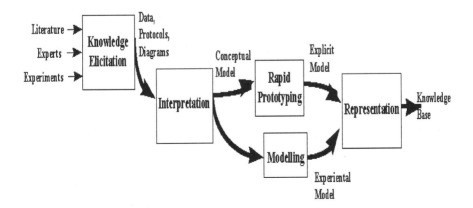

Figure 5.2 *Knowledge Engineering Process*

For design tasks a synthetic problem solving approach has to be used. One method for synthetic problem solving is KADS (Knowledge Acquisition and Design Structuring). It was developed in ESPRIT Project 1098 by Breuker and Wielinga (1987). KADS defines four stages of designing a knowledge base, of which the most important stage is the development of a conceptual model of expertise. At IWF a model-based knowledge acquisition method for design tasks was developed based on KADS (Göbler, 1992). This approach was used for the product structuring problem in DEFMAT.

5.3   Functional Specification of the Analysis Engine

As part of the development of the DEFMAT architecture, it was required to develop a generic engine that supports the problem solving requirements in DFM/A analysis. The overall system performance was assessed with respect to the delivery of functionalities specified in four demonstration scenarios (Figure 5.3) based on a Mobile Telephone design. The main types of analysis initially required to be supported by a single DFM system were:

a) Off-line DFA analysis of PCB assembly for the SMT assembly placement processes. The PCB assembly design is evaluated for placement problems with alternative assembly lines (i.e. equipment choices) to be checked against the same design.
b) On-line conceptual DFA analysis of the assembly of the PCB board into the telephone housing (the system helps the user to choose an appropriate joining method for the parts through generation of a ranked list of joining methods).
c) On-line detailed DFA analysis of the assembly of the housing. This involves detailed analysis of the actual joining features of the parts to be joined.
d) On-line DFM analysis of the detailed design of holes in the outer casing of the portable telephone. The hole-making methods currently available as options are drilling, boring, and milling. Cost and manufacturability criteria are used in assessing alternative manufacturing methods. Table 5.1 summarises those areas .

Table 5.1 *Coverage of DEFMAT Scope by Prototype System*

		Hole making	*PCB*	*Conceptual Assembly*	*Detailed Assembly*
Product	Machined Parts	X			
Domain	PCB		X		
	Telephone			X	X
Design	Conceptual Design			X	
Focus	Embodiment Design				X
	Detail Design	X	X		X
	Process Selection			X	
	Equipment Selection		X		X
	Engineering Change	X			
Mfg	Drilling	X			
Process	Tapping	X			
	SMT Placement		X		
	Wave soldering		X		
	Screw fastening			X	X
	Snap fit			X	X

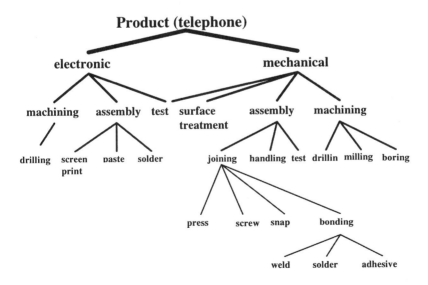

Figure 5.3 *Processes involved in Prototype Scenarios*

DFM/A analysis is complex, requiring significant information processing capability. Different forms of data models and design problems must be handled efficiently. A thorough analysis of the DFM/A requirements was performed to develop the specification of a suitable analysis engine to support the problem solving in DFM/A.

There are different expert systems shells, environments as well as general purpose CASE tools in the market. Mapping these capabilities to the DFM/A requirements allowed the selection of an implementation path for quick development of an industrial prototype. The range of design support functionality provided by the prototype is summarised in Table 5.2.

Table 5.2 *Design Support in Prototype System*

	Hole making	PCB	Conceptual Assembly	Detailed Assembly
Range of regimes	X	X	X	X
Selection of manufacturing resources knowledge bases	X	X		
Selection of product domain knowledge bases	X			
Interactive analysis	X		X	X
On-line analysis	X		X	X
Off-line analysis	X	X	X	X
Analysis with geometry	X	X		X
Analysis before geometric definition			X	
Provide design advice	X		X	X
Analysis of complete part	X	X	X	X
Analysis of single feature	X			X

5.4 DFM/A Methodology and Implementation

A study was conducted as part of the DEFMAT project to abstract the generic part of DFM/A methodologies. The purpose of the study was to define a generic set of problem solving requirements for DFM/A. The result of this study allowed the potential of integrating different problem solving paradigms to be used in the DFM/A domain. The optimisation of problem solving and data representation could then proceed.

In the development of a generic analysis engine for the use in the DFM/A domain, the particular decision characteristics needed to be carefully examined. The set of decisions and problem solving requirements were analysed to look for ways to achieve efficient use of problem solving techniques and thus their implementation.

5.4.1 DFM/A Problems

DFM/A problems can be generally divided into three main categories:

1. whether a feature can be made (Figure 5.4).

2. whether two features can be joined (Figure 5.5).

3. the selection of the best equipment for assembling a product (Figure 5.6).

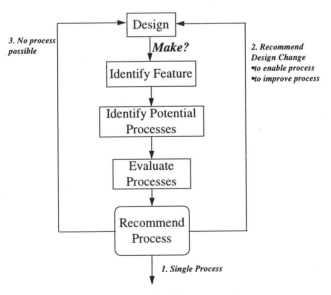

Figure 5.4 *Make Feature Analysis*

Effectively the DFM/A problem consists of two sub-problems. That of process and equipment selection and also provision of advice to the designer to improve the design to achieve "better" make or assemble characteristics. The two problems are interrelated and iterative. The DFM/A problem can start with an initial design to be made with any production means; or an established manufacturing facility constraining the freedom of design. In real life, the application always sits in between the two ends. These characteristics are captured in the Design Regime of DEFMAT. The above DFM/A analysis requirement can be generalised into several mechanisms of problem solving.

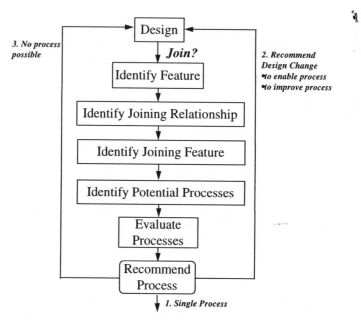

Figure 5.5 *Join Feature Analysis*

5.5 Problem Solving Paradigms

The five problem solving paradigms identified from the typical DFM/A analysis are:

1. Sequential Procedure
2. Algorithmic Evaluation
3. Constraint Satisfaction

4. Goal Searching
5. Ranking and Sorting

Sequential Procedure   This is effective if a known procedure or heuristics are to be applied. Procedure is an efficient form of processing as no solution plans are needed to be derived from rules. An example is the execution of particular DFM/A strategies as in the previous examples. In the implementation of this paradigm, the dynamic selection of procedure will enhance the flexibility of the tool.

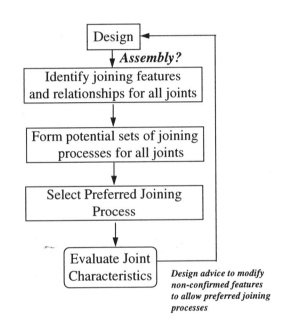

Figure 5.6 *Assembly Analysis*

Algorithmic Evaluation   In many operations, results in the form of a single value or vector is required with known and established formulae. This simple form of evaluation must be supported in any tool. An example is the evaluation of cost functions as a basis of selecting alternatives. In the implementation of algorithmic evaluation, dynamic loading of formulae parameters greatly enhances flexibility of application.

Constraint Satisfaction   In the selection part of the DFM/A problem, the typical starting point is the evaluation of a starting value against a set of constraints. Knowledge is represented as constraints, they are the factors that must be considered. The actual parameters within the constraints specify the bounds that must be satisfied. An example is the location of possible equipment for manufacture. Implementation consideration of this technique is critical to application development. A suitable satisfaction search path has significant impact on processing efficiency. Constraints are very suitable for object-type representation.

Goal searching   Many problems can be expressed as a single defined goal that does not require optimisation. Knowledge in goal searching is represented as relationships that form the tree to be searched. An example is the searching for joining relationship. In implementation, the structure and strategy of tree searching is crucial for efficiency. However, knowledge maintenance in the search tree is important but very difficult with a large search space, as the interrelationship of rules are not explicit.

Ranking and Sorting   Optimisation of choice is essentially the prioritisation of a set of alternatives. Multiple alternatives could be represented by a vector set of measures. Well established techniques exists for ranking and sorting lists. An example is the selection of best equipment from a list of candidates. Implementation of these techniques in n-tuple lists is well known.

5.5.1   Integration of Problem Solving Paradigms

The generic problem solving elements are seldom applied on their own. Most of the decisions involve a decomposition of the original problem into smaller sub-problems to be solved. An example is the problem of process selection. This goal is satisfied by the execution of a procedure to select processes. A set of process that the feature parameters could satisfy is generated. The cost function of each processes is evaluated with known cost functions. The selection is then completed by ranking the cost functions. To effectively support these decisions, the implementation of the problem solving engine requires a control to decompose and sequence the solution. Each problem solving mechanism has to be implemented for recurring and multiple calls.

5.6   Relationship between Product and Process Models

The problem solving mechanisms are implemented as objects that provide services to the main calling element of the DFM/A analysis. The key work is in the definition and implementation of design and manufacturing knowledge in the Product, Process and Manufacturing Resources Models.

In object-oriented form, the product and process models are represented in the form of objects. The interaction of objects is started by the DFM/A Control firing the appropriate trigger for analysis. This can be triggered in different ways depending on the type of analysis being performed (see Chapter 3), such as on-line or off-line analysis. This is responded to by the product feature in the product model which seeks appropriate processes. The exchange between the product and processes as defined by their methods involves the use of one or more problem solving mechanisms. These mechanisms are triggered by the appropriate messages from the DFM/A analysis control of the product and processes objects.

Manufacturing knowledge is captured in the Process and Manufacturing Resources Models. These models define the capabilities and characteristics of manufacturing resources. They are grouped according to their industrial domains and levels of abstraction as defined in the Design Regime. This allows selective loading of manufacturing knowledge base to focus on the design problem on hand.

The product model defines the product features and incorporated methods to associate with the appropriate manufacturing processes and equipment. These associations are completed with the execution of the DFM/A analysis. The Control Module governs the sequence of analysis as appropriate to the Design Regime and user requirements.

The process model is an important part of the DEFMAT architecture, as it represents the manufacturing knowledge which is fundamental to the making of DFM/A decisions. The scope of the Process Model encompasses the Equipment and Resource Models. The process model is a repository of manufacturing and assembly processes information and knowledge to perform DFM/A analysis. The process model is an information model, whose key design characteristics are its structure and interfaces to other modules. The model design criteria are the ease of use/retrieval of information; and the ease of maintaining the information.

5.6.1   Process Model for DFM/A Applications

It must be recognised that DFM/A analysis can be applied at different stages of the design life-cycle. The process model developed for the DEFMAT architecture must be capable to support the information requirement at the different stages of use.

A key consideration in the design of the process model is information uncertainty. In performing DFM/A analysis, the decisions are based on certain sets of assumptions on the processes and their capability. The accuracy of these assumptions dictates the validity of the analysis. Until the part is committed to manufacture, these assumptions are not reality. The earlier the DFM/A decisions are made, the more likely the assumptions will change. These changes can be in many forms. In a high technology area, process capability can be significantly improved in the time between the product design phase to the actual production phase. Material property expected at the design stage is subject to variation at actual production. This is particularly evident in rolled stock and short shelf-life chemicals. Depending on the status of manufacturing and subcontracting, alternative processes not considered at the design stage may become available at the production phase. Alternative processes are also important if the planned process could not be employed due to production capacity constraints. It is usually impossible to foresee the workload on manufacturing processes in the early stages of design. Market fluctuations in material cost can invalidate early selection of material and processes. For new processes, the stability and yield of the process may differ greatly from the assumptions made at the design phase.

To support DFM/A analysis subject to these uncertainties in information accuracy, the process model must be able to represent information at different levels of abstraction to be used for analysis at different phases of design.

A broad range of manufacturing processes is represented in the consortium. The DEFMAT architecture that can work with the richness of the processes in the consortium ensures a wide coverage of application domains. The challenge in the design of the representation is in the flexibility of knowledge and data representation to cater for the differences in requirements and characteristics in the different processes. This is in addition to the requirement to satisfy the time varying information uncertainty as discussed above.

5.6.2 Process Model

The object-oriented representation adopted by the consortium architecture make use of the class and inheritance structure to perform information abstraction of process knowledge.

Manufacturing processes and resources information are grouped into manufacturing process families as depicted in Figure 5.7. This allows selective loading of knowledge and data for analysis. This focusing on the application domain is crucial for processing efficiency and also knowledge maintenance. Knowledge maintenance is done by engineers with professional knowledge in the particular application field and concentrates on their particular expertise.

The issue of information uncertainty is addressed by the partitioning of manufacturing knowledge into three categories:

1. The broad process rules that contain general rules governing the use of the process. These general rules could be applied at the early phase of design.
2. Process constraints encapsulate the state-of-the-art capabilities or plant level capability of the process. This level of abstraction supports process selection at the later design phase. These first two levels are kept in the Process Model part of the DEFMAT architecture.
3. Equipment constraints detail the specific capabilities of the shop-level equipment and are most useful in supporting decisions at the production planning phase. They are kept in the Manufacturing Resources Model of the architecture. The use of process information depends on the command of the DFM/A control module that governs the DFM/A analysis and the product model methods.

Examples of the implementation of these process knowledge and rules in object form for a process and related manufacturing resource is illustrated in Figure 5.8 and Figure 5.9.

The process behaviour is defined as methods which will interact with messages being fired in the system. The methods used in the examples concerns the checking of process constraints that can satisfy particular product feature requirement and volunteer the process for the product feature.

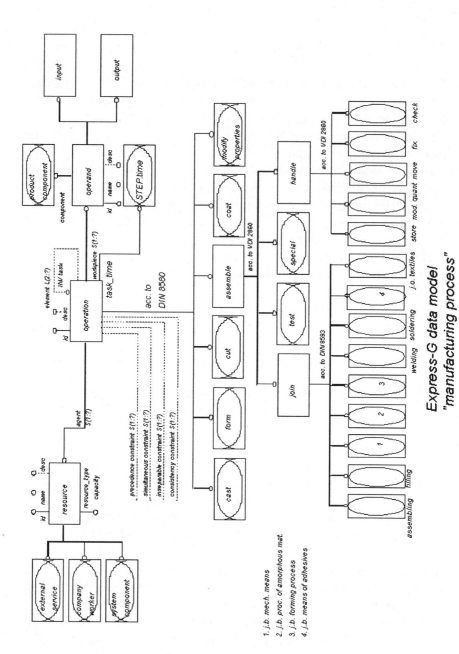

Figure 5.7 *Manufacturing Process Class Structure*

5.7   Triggering of Analysis

In order to achieve our final goal, the application of DFM/A rules to all or part of a product model, there must be a mechanism for triggering of the DFM/A rules. Returning to the different modes of analysis described in Section 5.3, we see that a generic system must be capable of supporting both on-line and off-line analysis.

```
CLASS :        Adhesive Joining
SUBCLASS:      Adhesive

CLASS :        Adhesive
SUBCLASS:      Anaerobic, Acrylic,.....
     ATTRIBUTES  :        Material [Steel, Aluminium....]
                          Service Temperature Min: -10 C
                          Service Temperature Max.: 120 C
                          Strength Min : 5 Kpa
                          Strength Max. : 15 Kpa
     METHOD:              "Joining Process?" message
                          Check Material
                          Service Temp Max.
                          Service Temp Min
                          Strength Max.
                          Strength Min
                          Flag messages to user

CLASS :        Anaerobic
     ATTRIBUTES  :        Cure Speed
                          Viscosity
                          Gap Fill
                          Equipment [Dispenser........]
```

Figure 5.8 *Process Object*

On-line analysis requires that direct communication exists between the CAD system, either via direct integration or using files to transfer information (see Chapter 4). Off-line analysis, as the term implies, involves the analysis of a product model read into the expert system using an interface file. On-line analysis may be further divided into two categories:

1. Nudging: the CAD Interface monitors the designers' actions and signals them to the DEFMAT Control Module. The Control Module uses the System Variables to determine if the design action should be analysed. System Variables (described more fully in the following chapter) are in effect the system configuration settings, deciding the appropriate design regime (hence which knowledge bases should be loaded) and which types of design action should trigger an analysis.
2. On-demand: the designer uses specially written CAD menu options to request analysis of a particular design feature.

CLASS:	Dispenser	
	ATTRIBUTE:	Needle clearance
		Needle height
		Needle angle
		Needle diameter
		Valve setting
		Valve pressure
		Robot speed
		Robot acceleration
		Robot tolerance
		Robot travel
	METHOD:	"Check dispenser" message
		Check Needle clearance
		Needle height
		Needle angle
		Needle diameter
		Valve setting
		Valve pressure
		Flag messages to user
	METHOD:	"Check robot" message
		Check Robot speed
		Robot acceleration
		Robot tolerance
		Robot travel
		Flag messages to user

Figure 5.9 *Manufacturing Resource Object*

The detailed DEFMAT architecture implementation required to support these types of analysis is described in the following Chapter 6. Here, we will demonstrate the triggering mechanism and linking of product and process and manufacturing resource (equipment) models for the addition of a hole-type feature to the CAD model.

5.7.1 Analysis Example

The blind hole feature shown in Figure 5.10 describes a standard blind hole, positioned using Cartesian co-ordinates. The standard parameters describe diameter (toleranced), depth (toleranced) and bottom angle. In addition the surface quality, cylindricity and remaining material at the bottom and side are specified for analysis.

The corresponding manufacturing feature is **hole_blind**. The typical manufacturing method for this type of hole is centre or counter sink drilling followed by blind drilling. The specified tolerances decide on the ease of manufacturing. The depth vs. diameter and the remaining material can be checked.

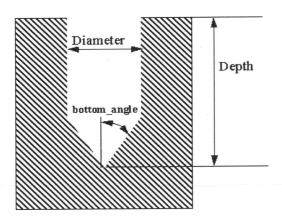

Figure 5.10 *Definition of blind hole feature*

The first step in adding the feature is the design selects CAD option "add feature" and then selects the feature from the list of user defined features in the CAD database, as shown in the Message Flow Diagrams (MFD) in Figure 5.11 and Figure 5.12.

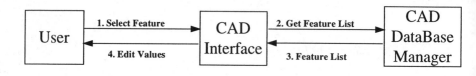

Figure 5.11 *Select Feature*

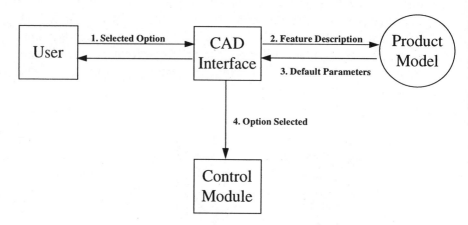

Figure 5.12 *Add Feature to Design*

The CAD Interface updates the Product Model in memory with the feature information. The CAD Interface is responsible for loading the default data from the product model in memory. The actual parameter values are filled in using the CAD functions, i.e. we do not copy a major part of the CAD functionality in the DFM system. Finally the CAD Interface sends the *"updated_feature"* message to the Control Module.

The next step is that the Control Module must determine, based on the system settings (System Variables), if an analysis should be triggered. IF so it sends the appropriate message (*analyse_single_feature*) to the Analysis Engine (Figure 5.13). In the meantime, due to the CAD Interface and the DFM system being separate processes (see Chapter 4), it is possible for the designer to continue to work, while the DFM analysis is taking place, potentially on a different computer. Either way, the DFM system will display its own user interface on the designer's machine. This provides detailed analysis feedback, without interfering with the CAD User Interface functionality, and process independence.

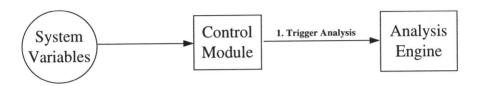

Figure 5.13 *Trigger Analysis*

The type of hole determines the type of processes and associated material removal equipment to be used. The process capability tables for various types of machining presented in Chapter 2 can all be expressed as DFM rules relating product or feature attributes to process constraints. An example of this type of rule is:

IF   hole_type = blind
THEN core drill

The tolerances associated with the hole dimensions as well as the surface finish required also dictate the process type. Examples of such rules linking feature attritutes to process capabilities are:

IF   hole_tol >= 0.002
AND   hole_tol <= 0.025
THEN drilling

IF   surface_finish >= 250
AND   surface_finish <= 100
THEN drilling

Process capability will also dictate allowed feature dimensions such as maximum hole diameter or the hole length:diameter ratio, as expressed in the following rules:

IF   hole_dia >= 150
AND   hole_dia <= 0.5
THEN twist drill

IF   hole_dia >= 150
AND   hole_dia <= 25
THEN spade drill

IF   hole_L/D_ratio <= 50
THEN twist drill

IF   hole_L/D_ratio <= 100
THEN spade drill

Once a process is chosen, features can later be altered by changes in major or minor dimensions, tolerances, surface finish or even material type. Such changes may also trigger analysis, and the re-checking of feature attributes may trigger rules such as:

IF   hole_dia> Proc.dia.max
OR   hole_dia> Proc.dia.min
THEN Message ("hole diameter exceeds maximum allowed for process selected")

During the analysis the Analysis Engine will present the user with a list of possible processes, ranked in order of preference, if there is more than one suitable. Once the user has chosen the preferred process, it may also be necessary to provide design advice relating to the process selected. For example for drilling there will be rules associated with tool (drill) clearance.

5.7.2   Design Advice for Machining Example

There has to be clearance between the drill and the nearest object to allow for swarf removal. The amount of swarf generated depends to a large extent on the drill angles and its diameter. It is fair to say that the clearance required is directly proportional to the drill diameter. Hence

$$xd = y$$
$$x = y/d$$

where:

x = swarf clearance required
$y = w-d$ (actual clearance)
w = distance between hole centre and object
d = drill diameter

Condition: In order for sufficient clearance to exist

$$x < y$$

finally
$$a = y - x$$

where a is the extra distance required between the hole and the object. This type of design advice can then be expressed as:

IF $y < x$
THEN Message("Move hole away from object by distance a")

The Analysis Engine, having applied the appropriate product and process knowledge sends an *analysis_result* message to the Control Module containing the analysis feedback for the user (Figure 5.14). This triggers the Control Module to send a *show_analysis_result* message to the DEFMAT User Interface. This message contains the actual analysis output, including information on the type of output (which determines how it is to be displayed), and whether user feedback is required. For example, it may be a list of processes, ranked in order of suitability, from which the user must choose before the analysis can continue to perform more detailed DFM relating to equipment selection for the chosen process.

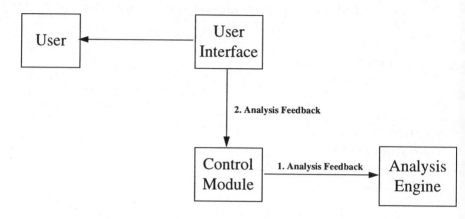

Figure 5.14 *Analysis Feedback*

Depending on the user feedback, the Analysis Engine may request the Control Module to update the DEFMAT Product Model with the process information relating to the feature (Figure 5.15).

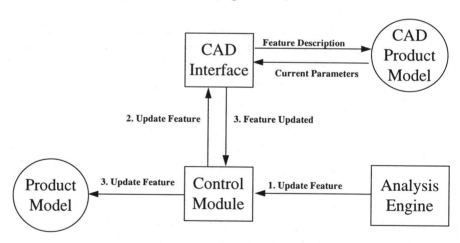

Figure 5.15 *Update Feature*

The update feature function is identical to the add feature function. Instead of using the default values, the reference to the CAD geometry is given. Using the standard CAD functions, the feature is edited and the new parameters are stored by CI in the product model. The CM receives a notification that a feature was updated, and can react according to the System Variables settings.

6. Systems Implementation

At the start of the DEFMAT project, it was essential that an initial skeletal structure was laid down for a system architecture in order to provide a basis for the commencement of work on four prototypes. This initial structure was the result of experience and intuition rather than a purely analytical approach. The four initial research prototypes were based on this initial architecture (Figure 6.1), which was used to identify the major DFM architecture components and interfaces.

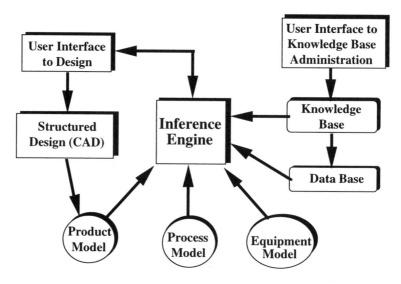

Figure 6.1 *DFM System Architecture Template*

In order to show the convergence towards the final architecture and system, it is instructive to review briefly these research prototype systems, each of which was developed for a particular industrial design and manufacturing regime. These range from surface mount assembly of printed circuit boards, to complex machining. The experience gained in the specification and development of these prototype systems provided the basis for the specification of the generic DEFMAT architecture.

6.1 System for Design for PCB Assembly

The prototype developed by CIMRU and Digital (Figure 6.2) was concerned with the Digital Surface Mount PCB assembly process. The design data was available from two sources: an interface with the Digital proprietary design data standard - ADS; and an interface with the 3D parametric feature-based modelling system, Pro/Engineer. Both sources are interfaced with a Product Model defined using the STEP/Express standard. Process engineers can store DFM knowledge in the form of rules in a relational database via an X-windows user-interface, with active links to a model of current manufacturing resources. The results of the DFA analysis are available to the designer via another X-windows interface. The basic programming tools of C, SQL and X-windows were used exclusively in the development.

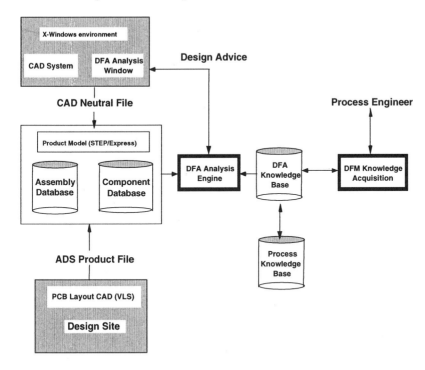

Figure 6.2 *System for Design for SMT Assembly*

6.2   System for Design for Small Parts Assembly

The prototype developed by CIM Institute (Cranfield Institute of Technology) was based on the assembly activities of a UK producer of small electro-mechanical products, with the emphasis being on adhesive joining techniques. The system (Figure 6.3) interacts with the designer during design activity on the CATIA CAD system. Knowledge on available joining techniques and current DFA rules are implemented using a combination of C and SQL. Some knowledge is implemented using a modular programming approach. Other languages used are FORTRAN and IUA (CATIA-specific application programming interface language).

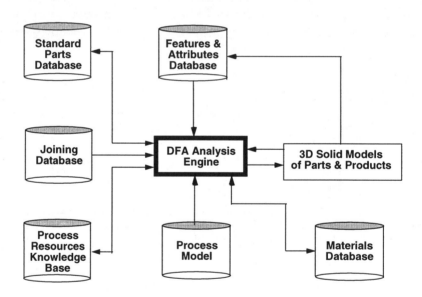

Figure 6.3 *System for Design for Small Parts Assembly*

6.3   System for Design for Mechanical Assembly

The prototype developed by IWF and AEG is based on the operations of the AEG mobile communications division in the assembly of their mobile telephone units. The prototype system (Figure 6.4) consists of the CAD modeller ACIS, the expert system tool ProKappa and the relational database ORACLE. The user interface is realised using X-windows. The integrated product and process model was developed using the STEP/Express language. The interface between the knowledge base and database automatically creates and maintains relations from the object-oriented description in ProKappa. A first module of the DFA analysis engine was realised which estimates assembly costs based on a dialogue with the designer and additional cost information from the database.

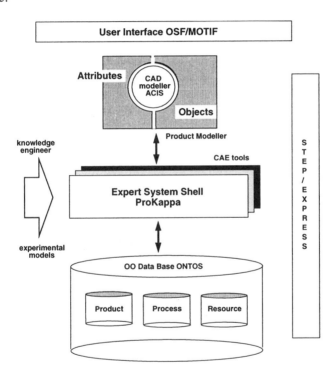

Figure 6.4 *System for Design for Mechanical Assembly*

6.4 System for Design For Machining

The objective of WTCM was to build a prototype for cost minimisation, focusing on the turning, milling and grinding production processes involved in making shafts and rotors for air compressors. The prototype (Figure 6.5) was made in collaboration with Atlas Copco, Antwerp. The

WTCM prototype was based initially on the design and manufacture of rotor shafts, with further research into the overall design methodology for complete compressor units. A first working prototype was completed using the Nexpert™ expert system and the Applicon CAD system, focusing on the design with complex form features and integration of the design with the standard parts database. The BRITE MODESTI interface was used for the integration of the expert system with the CAD system, and the link with the expert system written in C. A "question and answer" system is used as the user interface.

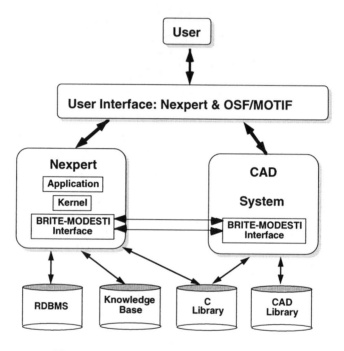

Figure 6.5 *System for Design for Machining*

6.5   The Generic Architecture Operational Aspects

The generic architecture (Figure 6.6) must incorporate certain elements such as a design system but the disposition of these elements and associated modules enable the final goals to be realised. It is also essential that within the information flows in the architecture, distinction must be made between

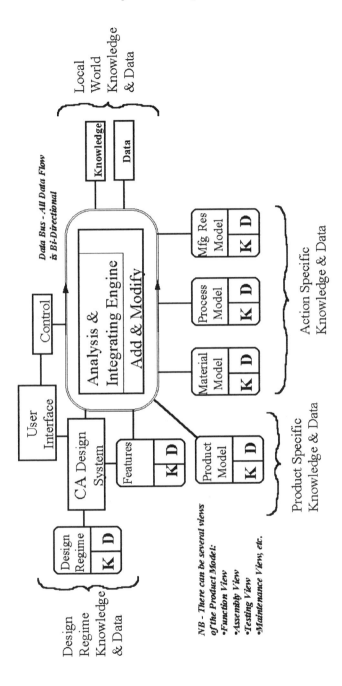

Figure 6.6 *Conceptual Generic Architecture*

knowledge and data. Knowledge may describe a process but data removes the genericity and turns the process into a specific action at a specific factory.

Because of the quantity and types of information flow that may take place in the architecture, the concept of an information highway or data bus was employed with all modules and system processes being able to communicate via this data bus.

The user interacts with a unified user interface. By unified, the interface appears to be a single entry port into the system but in fact deals with two aspects; direct entry into the Computer Aided Design system and direct entry into the control system. The Computer Aided Design System may be any such system. Naturally the type of system used reflects in the richness required from the control system.

The control system receives instructions either from the user interface or from other system modules and acts accordingly. The knowledge and data attached to the control system are to enable it to be adjusted with respect to the environment in which it will be operating.

The features store or database contains all types of features. These features may have the following forms: totally fixed geometry, fixed geometrical modifications, parameterised geometry. By totally fixed geometry is meant either "blocks" of component parts that may be incorporated into new components or complete parts, such as proprietary fixings, that cannot be described parametrically. Fixed geometrical modifications are typified by holes on a pitch circle diameter. Parameterised geometry is geometry which can be described in terms of relationships and variables such that a wide range of geometry can result from the input of data.

The product model is the total representation of all aspects of the product and it may be presented in a number of views such as the function view, the assembly view, the testing view, the maintenance view, etc. Product models have a very long life and under go changes during that life, all of which must be recorded. STEP/Express methods are employed for its description but do not satisfy fully the total requirements of the product model.

The material database contains all of the aspects of knowledge and data relating to the materials that may be employed to realise the product. The processes model similarly contains information on processes that may be employed, whereas the manufacturing resources relate to those resources that are available for the production of the product. A process, for example may be deep hole drilling where the knowledge relates to the best materials and problems that may be encountered. The manufacturing resources may indicate that such work cannot be performed in-house and indicate cost penalties.

In order to handle the rules and conflicts and to monitor the design process, the Analysis and Integrating Engine is central to the architecture. The knowledge and data are required to be transformed into a form recognisable by the Analysis and Integrating Engine.

If the architecture is examined it can be seen that it is also divided into zones - the design regime knowledge and data, the product specific / action specific and local world zones. Design for manufacture is one small part of the design domain. Because of the breadth and depth of design, it is usually practised in specific areas, thus engineers talk of mechanical design, electrical design, industrial design and so on. Each of these areas is usually subdivided into domains such as the design of pumps, compressors, digital circuits, rotating mechanical machines. In each of these areas there exists specialist knowledge that may be manifested as analysis or synthesis tools, rules from certification societies, legislation or from within the company. Certain aspects of design may exist in different areas, for example, bearing design. The parameters and degree of acceptability may differ between these areas due to operating constraints, life, quality, cost and many other constraints. Thus those factors that are intrinsic to a design area in order to achieve the design intent are designated in the Design Regime of the DEFMAT architecture . Generally the rules in this regime take precedence over design for manufacture rules if the design intent and performance parameters will be violated.

In performing his tasks the designer is working with levels of abstraction. It is of paramount importance that these representations fulfil the perceived needs and that the problems solved are the real problems and not simplified substitutes. Many levels of abstraction have to be supported by the design regime.

Complex designs have to be subdivided, generally across boundaries of weak interaction or using a lumped parameter approach. This is obvious when considering the design of products such as internal combustion engines or aircraft. Where the total approach is virtually impossible and the product is designed by splitting it into subsystems, Design for Manufacture and Assembly imposes constraints across these boundaries in terms of dimensioning and assembly.

In approaching a design, a designer may use several iconic forms, each of which is related to a stage in design. It is only at these final stages of component design, that design for manufacture comes into play. In electronic, electrical, process plant and other engineering disciplines, iconic forms exist but have different representations and different information attached to them. In electronic engineering at the point in design where the net list becomes routed, (from the so called "rats nest" to the initial board layout) there should be stronger links with design for manufacture.

As may be deduced the design regime is very rich and provides scope for further development. The interaction between design rule sets and manufacturing rule sets may become complex as the influence of manufacturing shifts to the upstream processes of design.

This shift is already considered in the modes of operation for the DEFMAT system. At the plant level one is dealing with broad design concepts. This is the basic process of sizing a manufacturing unit. It must not be forgotten that design can be a recursive process for a design system is used to design the plant resulting from the sizing process. Similarly a design system can design a design system when it is given the appropriate inputs.

Many aspects of the design regime may be expressed as graphical networks. The design regime must be able to act on these nets and associated rules to provide a convergence, under the control of the designer, to the specific design problem.

Thus the design regime is that set of knowledge, data and rules expressed in a wide variety of forms. These forms range from computer programmes (finite element, synthesis, simulation etc.), data files, reports, knowledge bases, test results etc.. These forms of information are utilised by the designer and CAD system to realise the various aspects of the product design. Design for manufacture is an additional influence upon these actions that may modify the design by imposing additional constraints.

Output from the design regime is also added to the product data model and new designs may reference existing product models. Thus the design regime provides the substance that lies beneath the images, technical drawings, pictorial representations and schematics that represent a product.

Another aspect of the foundation upon which the architecture is constructed is that of object technology. The object-oriented approach is utilised for the databases, the software construction and the AI engine. Because of the nature of the present AI tools, one of the functions of the control module is to recognise when the design engine requires different rule sets to be loaded into the AI engine.

The DEFMAT project has demonstrated the viability of the architecture and relies upon it implicitly. The architecture also has scope for extending the work into other areas such as production planning and *Design for X* where *X* is the dominant constraint, such as cost, life, weight, etc.. The architecture has also drawn attention to the central role of the design regime and the effort required to integrate design performance tools.

6.6 Architecture Realisation

The architecture may be implemented within a single workstation or over a number of system elements. The user interface protects the users, who whatever path is taken, see the system as a single entity. The industrial prototype realisation of the architecture is shown in Figure 6.7.

6.6.1 User Interface

All user interaction with the DEFMAT system is through the User Interface (UI) (see Figure 6.8). It also provides any windows necessary for user input and output during analysis. It interacts directly only with the Control Module (CM). The UI is implemented using OSF/Motif and C. Further versions could be implemented using even more flexible windows development software which would generate platform specific code from the generic windows code.

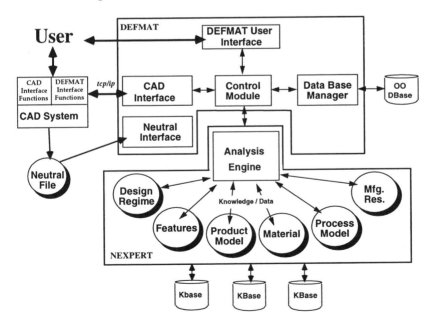

Figure 6.7 *Architecture Realisation*

This module incorporates the window manager of the DEFMAT system and all functionality relating to windows. When the DEFMAT software is activated, its first call is to the CM to initialise Nexpert™ and perform whatever actions are defined for system start-up (e.g. loading of the Control Module Knowledge Base).

The first input to the DEFMAT system is via a password window to identify the user and establish his authority. This allows the system to distinguish between users with different privileges, in the case where for example, only authorised personnel may access the DFM/A knowledge bases.

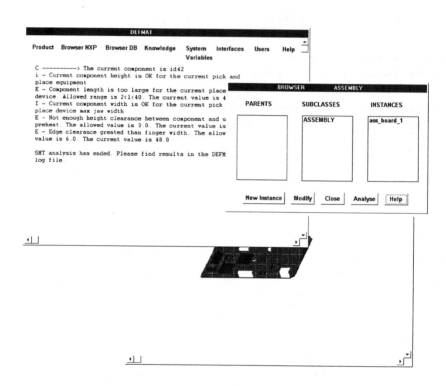

Figure 6.8 *DEFMAT User Interface Main Window*

The main DEFMAT environment is then available to the user. The objective in implementing the UI was to allow the user as full as possible access to the concepts embodied in the architecture and implemented in the data structures and software modules. The UI itself is not intended to be intelligent - all decision making regarding the DEFMAT environment is made by the CM. The UI literally provides a "window" onto DEFMAT. The main UI window provides functionality under the main headings of Product, Browser, Interfaces, Knowledge, System Variables and Users.

6.6.2 Product Options

This main menu item provides the user with options to create, modify, analyse and save products.

6.6.3 Browser

Two Browser main menu items are implemented - one for browsing Nexpert™ (i.e. data in active memory, accessible to Nexpert™), and one for accessing long-term storage in the ONTOS object-oriented database. Both items provide options to browse up and down through the class hierarchies of products, processes, parts, joints, materials, etc.. The user may create new instances of classes or modify existing instances (Figure 6.9). The creation of new classes is restricted to users with manager status, who must use the Knowledge main menu item to enter the knowledge processing environment (Nexpert™) in order to modify class definitions and rules. The Database Browser allows the user to load objects from the database into active memory, where they may or may not be linked with other objects or incorporated into products. The Nexpert™ Browser provides the user with an "analyse" button. When an object is selected and the user selects "analyse" the CM requests the Analysis Engine to perform an analysis of that object. The type of analysis performed depends on the design regime selected through the System Variables main menu item.

6.6.4 System Variables

In order to customise the DEFMAT environment, this main menu item provides a window in which the user may edit the current values of the system variables. Any number of new system variables (see Table 6.1 may be added by a user with privilege to modify the CM knowledge base, where they reside. New system variables may be subsequently used in the control rules, or by the Analysis Engine to control analysis.

When the values of the system variables are changed the CM automatically checks to see what knowledge bases must be loaded or unloaded.

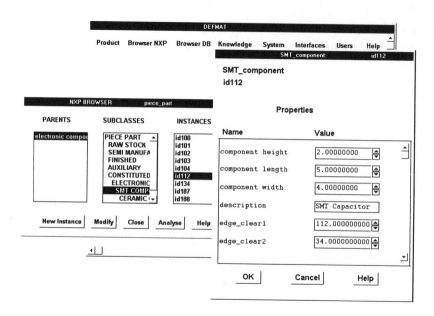

Figure 6.9 *Browser Window*

The other options which the UI provides are the Interfaces option, where the user can:

- choose and open a particular CAD system from the list of available CAD systems (in the first implementation the Pro/Engineer interface is implemented).
- import a Pro/Neutral file (in this implementation it is used to import the AEG PCB assembly from Pro/Engineer to the DEFMAT product model).

- enter the Nexpert™ environment (if the user has manager privileges) in order to modify or create class definitions or to modify or create rules.
- enter the User option, where the user can create or modify users and their passwords, delete users and assign privileges.

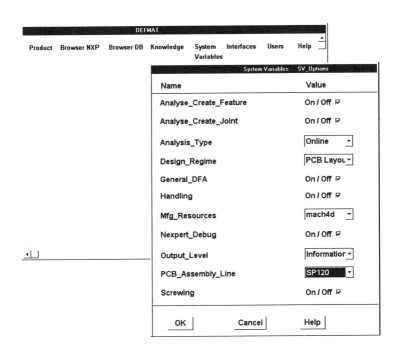

Figure 6.10 *Changing the System Variables*

6.7 Control Module

The control module is implemented as a combination of code and rules. The code is implemented as C++, and provides an array of basic functionalities for accessing Nexpert™, the database and the User Interface. It also acts as a bridge between the Analysis Engine and the CAD Interface as well as the User Interface. The bridge between the Analysis Engine (programmed in Nexpert™) and the CAD Interface and User Interface is achieved by special handler functions, such that the

User Interface can be called directly from within Nexpert™ rules. The CM knowledge base contains:

1. classes necessary to model all the Design Regimes and System Variables
2. rules to link knowledge bases with Design Regimes and some System Variables

When the system variables are changed the control module performs its own analysis using its knowledge base to determine what knowledge bases should be loaded or unloaded. For example, if the design regime is changed to "Detailed Design", the following control rule is fired, loading the correct process and manufacturing resources knowledge bases, and unloading any unwanted knowledge bases (stored as *.kb files) which may be present.

IF

> System_Variable_Current.Design_Regime IS "Detailed Design"

THEN

HYPOTHESIS Change_System_Variable **TRUE**

AND

LoadKB	a1_matl.kb	LEVEL=ENABLE;
LoadKB	a2_rules.kb	LEVEL=ENABLE;
LoadKB	a2_screw.kb	LEVEL=ENABLE;
LoadKB	a2_grip.kb	LEVEL=ENABLE;
LoadKB	a2_s_eq.kb	LEVEL=ENABLE;
UnloadKB	pb_rules.kb	LEVEL=DISABLESTRONG
UnloadKB	a1_rules.kb	LEVEL=DISABLESTRONG
UnloadKB	mc_rules.kb	LEVEL=DISABLESTRONG

6.8   Knowledge and Data Maintenance

The knowledge engineering mode requires access to all data in the system (knowledge and numerical). This editing is done in the expert system's standard knowledge and data editor. Similarly, data can be edited in ONTOS. The knowledge engineer has full control of the modules within a knowledge engineering session.

Table 6.1 *System Variables*

System Variable	Possible Values	Explanation
Analyse_Create_Feature	ON/OFF	Feature creation triggers analysis
Analyse_Create_Joint	ON/OFF	Joint creation triggers analysis
Output_Level	Error, Information	Level of DFM/A output information required
Analysis_Type	Online, Off-line, Nudging	
Design_Regime	Conceptual Design Detailed Design PCB Layout Machining	Type of Design Regime
General_DFA	ON/OFF	Perform General DFA analysis
Nexpert™_Debug	ON/OFF	Open Nexpert™ for debugging purposes during analysis
Handling	ON, OFF	Perform handling analysis
Mfg_Resources	mach4d mach5d	4D or 5D machining resources available
PCB_Assembly_Line	HS180 SP120	The two PCB assembly lines available
Screwing	ON/OFF	Perform screwing analysis

Knowledge maintenance is facilitated by the structured way in which it has been formed. Figure 6.11 shows an object structure widow, similar to how it would appear in an expert system session. As the window image indicates, models are developed hierarchically. Hence, maintenance could mean simply adding more subclasses to existing classes to represent more features or functionality of products or equipment.

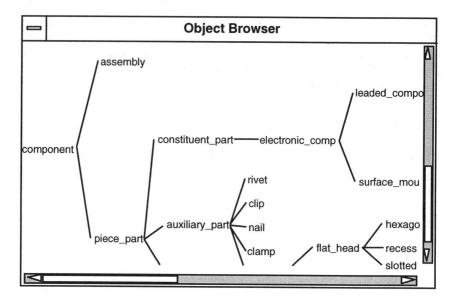

Figure 6.11 *Object Structure in Expert System Window*

The classes and objects stored in the current DEFMAT prototype are the result of the information gleaned from work at the 4 test sites which were used for the project. This information was used as a constant reference to adequately develop the system. If DEFMAT were implemented in a new design / manufacturing environment this would possibly occur in three phases:

1. The current model classes would need to be studied against the plant's actual product range and facilities to determine what alterations are necessary to suit the plant. (Even though the currently implemented classes are general across a broad range of manufacture/assembly, there will at least be special additions necessary).

2. All relevant values for the attributes of the Process, Manufacturing Resource and Material Models would be assigned.

3. All rules would need to be added by applying knowledge gleaned from interviews with manufacturing engineers, workers etc.. This process will be on-going.

The designer should now have a rich, up-to-date repository of manufacturing and assembly process information and knowledge to perform DFM analysis on Product Models that are generated through the CAD system.

6.9 Process Model and Manufacturing Resources Models

6.9.1 Class Definitions

The process model structure is defined in the "class.kb" file for all the scenarios. The conceptual process model was developed for a wide range of industries. Detail process capability were collected from AEG and Atlas Copco and modelled into the relevant process model files. The process model structure is loaded into the DEFMAT system all the time. Detail process constraints are loaded when necessary and unloaded to reduce the dynamic memory requirement of the running system. The top level definition of the process model is illustrated below.

```
CLASS NAME:   process
SUBCLASSES:   manufacturing
PROPERTIES:   name
              resource

CLASS NAME:   manufacturing
SUBCLASSES:   assembling
              machining
PROPERTIES:   name
              resource

CLASS NAME:   assembling
SUBCLASSES:   joining
PROPERTIES:   name
              resource
```

The class and object structure of the manufacturing resources model effectively matches that of the process models. It is defined to contain more detailed information about the capability of a specific plant and its equipment. Plant information is loaded dynamically according to the need of analysis. The high level definition of the manufacturing resources model is listed.

CLASS NAME: equipment
SUBCLASSES: PCB_equipment
            handling_equipment
            modifying_quantities_equipment

CLASS NAME: PCB_equipment
SUBCLASSES: PCB_wave_soldering_equipment
            PCB_pick_and_place_device

CLASS NAME: handling_equipment
SUBCLASSES: main_purpose_handling_equipment

CLASS NAME : main_purpose_handling_equipment
SUBCLASSES: buffering_equipment
            moving_equipment
            fixing_equipment
            checking_equipment

CLASS NAME: modifying_quantities_equipment
SUBCLASSES: screwing_equipment

CLASS NAME: fixing_equipment
SUBCLASSES: fixture
            jigs
            gripper
PROPERTIES: cad_file
            control
            description
            name
            path

6.9.2   Process Models

For the Conceptual Joint Design scenario, the process model is a big reference table of the process capability. The process model layout is effectively a two-dimensional matrix with possible processes as the row labels and the requirement values as the column values. The advantage of inheritance is used so that the values of children in a class is defined only if they are different from the parent. Otherwise, the value defined by the parent is used for all the children processes.

CLASS NAME   joining
SUBCLASSES: j_b_means_of_adhesives
                    j_b_mechanical_means
                    soldering
                    welding
                    j_b_assembling
                    j_of_textiles
                    j_b_processing_of_amorphous_materials
                    j_b_forming_processes
PROPERTIES (sample):
                    attenuation
                    cost_of_use
                    degree_of_automation
                    degree_of_difficulty
                    dynamic_load_bearing_capacity
                    static_load_bearing_capacity
                    electric_conductivity
                    fluid_impermeability
                    gas_impermeability
                    manufacturability
                    manufacturing_cost
                    recyclability
                    removability
                    repairability
                    required_auxiliary_material
                    required_auxiliary_parts
                    required_force
                    required_tools

In the detail joint design scenario, the focus is on the equipment model. A selection of grippers, screwing equipment and screws are available for the user to select. The class structure of these are defined in the common class definition and the equipment model files contains the value of geometry and constraints.

The machining scenario defines the general process-related values in the process model file. Plant-specific constraints are loaded through the user selected equipment model. The top level class is machining.

CLASS NAME: machining
SUBCLASSES: drilling
             grinding
             centring
             reaming
             threading
PROPERTIES: chk_mfg
             mfg_cost
             name
             resource

CLASS NAME: drilling
SUBCLASSES: drilling_surface
             drilling_counter_bore
PROPERTIES: chk_mfg
             mfg_cost
             name
             resource

CLASS NAME: drilling_counter_bore
PROPERTIES (sample):
             chk_mfg
             clamping_height_max
             clamping_height_min
             clamping_length_max
             clamping_length_min
             clamping_width_max
             clamping_width_min
             distance_to_bottom
             distance_to_edge
             l_d_max
             l_d_min
             mfg_cost
             name
             resource

surf_quality
tol_dial
tol_length
top_angle_max
top_angle_min

The equipment model is held in two files for the user to select a four axis machining or five axis machining. Similarly, the PCB scenario allow the user the choice of two different manufacturing lines. The properties for analysis is defined in the common class structure.

CLASS NAME: PCB_pick_and_place_device
PROPERTIES: allowable_above_board_protrusion
max_conveyor_height
max_jaw_length
max_jaw_width
max_pp_conveyor_width
min_conveyor_height
min_jaw_size
min_pp_conveyor_width
PTH_side_clearance
SMT_side_clearance

CLASS NAME: PCB_wave_soldering_equipment
PROPERTIES: conveyor_finger_width
distance_to_fluxer
distance_to_lower_preheat
distance_to_upper_preheat
max_ws_conveyor_width
min_ws_conveyor_width

In the Nexpert™ implementation, process data can be written and retrieved in tab delimited form which allows the knowledge engineer to prepare process data off-line using common spreadsheets or database applications. The DEFMAT process model architecture is successfully used in the industrial prototype to support a wide range of design for manufacture analysis.

6.10   Analysis Engine

The Analysis Engine in DEFMAT is developed as a generic problem solving engine for design for manufacture analysis. Its functionality is demonstrated in the industrial prototype performing analysis in four very different design for manufacture problems. The implementation aspects are discussed in this section.

The Analysis Engine is actually implemented as a sub-process of the main DEFMAT process - this allows the user to continue to use the DEFMAT environment during analysis. This is especially important when the user must make choices during analysis, and may wish to consult the Browser first.

6.10.1   Analysis in the demonstration scenarios

The DEFMAT prototype demonstration consists of four industrial scenarios. They are based on the AEG mobile telephone. The capability of the knowledge rules implemented is much more generic and powerful and could be used with a much bigger range of products. The four scenarios used to demonstrate the DEFMAT architecture are :

1.   conceptual joint design support
2.   detail product joining analysis
3.   machining process selection and evaluation
4.   PCB component layout checking

Conceptual Joint Design Support

The conceptual design of a joint relates to the structural layout of the product assembly and is a major decision in any product design. In the specification of a joint, the joining requirements and the selection of the joining process define the guidelines for the detailed design of the joint geometry and join features. In the conceptual design process, the designer is specifying the functional requirements of the product. The DFM support provided in this scenario supports the designer with the characteristics of known manufacturing processes. This allows the designer to consider the ease of manufacture as well as the pure functional aspect of design. The trade-off results in a more cost effective design of the product.

The analysis in the conceptual design scenario is a highly interactive calculator. There are currently 25 different requirements defined for analysis (Table 6.2).

Table 6.2 *Joining Analysis Requirements*

attenuation	overload toleration
centrability	quality testability
corrosion resistance	recyclability
degree of automation	removability
dynamic load bearing capacity	repairability
electric conductivity	reusability
fluid impermeability	shape
gas impermeability	static load bearing capacity
high frequency impermeability	temperature resistance
maintainability	thermal conductivity
manufacturability	variety of design
material efficiency	variety of load
operational safety	

The designer can specify the degree of significance for these requirements. A simple Yes~No, High~Medium~Low scale is used in this conceptual design phase analysis. Exact specification of values is part of the analysis process in the next design stage. Also, the information uncertainty of the corresponding capabilities of the processes is insufficient to pinpoint on exact values. In addition to the specification of joint requirements, this analysis allows deficiencies in current process capability to be highlighted and development could be done before the start of production.

The sequence of analysis consists of multiple loops of evaluation. After the designer specifies the joint for analysis, he can select the list of requirements that is important for this joint. For the value that he specified for each requirement, the analysis computes the processes that can satisfy his requirement specification and the processes that have to be rejected. The designer can go through the analysis as many times as he wishes until he is satisfied with the specification and knows the consequences in processes selection. This is repeated for all the specified requirements. Ranking values are applied to the possible processes based on:

1. auxiliary material
2. auxiliary part
3. manufacturing cost
4. material efficiency
5. operational safety
6. recyclability
7. removability

After the selected process is updated in the joint database for future reference and the log of the analysis is also saved. The analysis is performed at the early conceptual phase of design before the geometry of the parts are defined. This overcomes one limitation of most commercial DFM/A systems, which can only be used after assembly and process plans are defined.

Detailed Joining Analysis

The detailed joint analysis analyses three aspects of the product for joining. Three analysis types are defined and the designer selects the analysis he wants from the DEFMAT System Variables. A general design analysis checks that the combined weight of components are within specified product weight of the product. The general check also checks the elasticity of the part material and warns the designer of any material that is floppy and could be difficult to handle.

The handling analysis checks all the parts for the ease of automated robotics gripping. The analysis helps the designer to select the robotics gripper in the manufacturing resources database. Through-holes of diameter between 1 to 3 mm on the part are automatically found. Their geometric position is analysed and the three holes with maximum separation and most suitable for use as gripping holes are selected. The distance between holes are matched to the robotics grippers selected by the designer.

The screwing analysis helps the designer to select the screwing equipment available in the manufacturing resource database. The analysis evaluates the geometric position of the screwing hole and dome features on the parts and guides the designer in selecting the suitable screw for joining the two components.

The analysis is performed with the CAD interface running. The designer can take the design advice from the DEFMAT analysis log and modify the design in the CAD system.

Machining Process Selection And Evaluation

The machining process selection analysis evaluates the machining features on piece parts. The current demonstration incorporates only the analysis of holes and associated features. The automatic features extraction capability can identify the following feature types :

1. simple through-holes
2. simple blind holes
3. threaded holes
4. stepped holes
5. tapered holes
6. countersinks
7. counterbores
8. compound holes with a combination of the above
9. domes

These design features are mapped to the corresponding manufacturing features that have to be combined to machine these features.

The designer/process planner can select different manufacturing plants with their particular process and equipment capability. This information is held in the corresponding process and manufacturing resource database. Once this is set in the DEFMAT System Variables, the analysis is performed specifically for this plant's equipment. The analysis provides advice on design features on the part that cannot be machined in the plant and also calculates the cost of machining. This analysis is a first step to detail operations planning. The design advice is maintained in the DEFMAT analysis log for future reference. The designer can interact with the CAD system to implement the advice of analysis.

PCB Component Layout Checking

The PCB scenario addresses off-line analysis to find any potential layout problems in PCB assemblies for any manufacturing line. The designer selects the line for analysis in the DEFMAT System Variables. The capability and constraints of the line are defined in the corresponding resource databases. The PCB product information is loaded from the electronic product neutral interface once the user selects a part.

The dimensions of the complete PCB is checked for the fit in the different pick and place devices, placement and flux soldering devices. If through-hole components are on the board, then additional checking for the wave soldering process will be done. The board is checked for edge clearance for robotic pick and place, and gripping on the conveyor to go through the soldering devices. For each component, the combined height of the board with components is checked for height clearance through all the process stages. The minimum distance between components is checked to ensure the placement device in the line can cope with the requirements. Each component is checked with all its neighbours. Any problems are flagged to the DEFMAT analysis log. A typical board contains more than 60 components and can be over 100s. The analysis results could be used for the selection of PCB assembly lines and highlighting areas for product and process improvement.

6.10.2   Analysis Engine Architecture Aspects

The Analysis Engine is designed to conform to the DEFMAT architecture, to deliver total flexibility of application and implementation. This is achieved through the use of object-oriented representation in product and process models and the separation of knowledge and data in implementation.

The Nexpert™ Version 2.0 environment was selected as the implementation platform of the consortium. This is a stable environment for the consortium to build and integrate additional functionality. Object-oriented representation is used for the definition of objects and their properties. The definition of the behaviour and methods of objects are in the form of rules. The inheritance, data abstraction and information hiding of the object paradigm is implemented. Property type, naming and reference is explicit. These explicit forms are the common definition for the development of the corresponding data representation in the ONTOS database, the CAD interface and the user interface. Rules are kept independent of the objects and reference properties of object in the analysis.

To conform to the DEFMAT architecture, separation of knowledge and data is practised through putting the data and rules into separate knowledge base files (Figure 6.12). The knowledge base loading and unloading features of Nexpert™ is exploited to support user selection of processes and manufacturing resources. It is also used to unload unnecessary rule bases. To achieve flexibility of implementation, the definition of classes and the data values are separated according to the modules in the original architecture (Table 6.3).

Table 6.3 *Class Storage Structure*

	Conceptual	*Detailed*	*Machining*	*PCB*
Process model	a1_procm.kb	a2_procm.kb	mc_procm.kb	pb_procm.kb
Analysis rules	a1_rules.kb	a2_rules.kb	mc_rules.kb	pb_rules.kb
Mfg Resources		a2_grip.kb a2_s_eq.kb a2_screw.kb	mach4d.kb mach5d.kb	

The class definition file contains all the definition of class for all the scenarios. This represents the product, process and material model structure implementation in the Analysis Engine. The material model contains material data relevant to the AEG mobile telephone application and is common for all scenarios. The process model and manufacturing resources model files effectively contain just the data slot values for the particular analysis. The advantage of this implementation is the flexibility in changing analysis focus. For the processes currently defined, any product defined by the user could be analysed to evaluate its suitability for manufacture in the particular plant. Alternatively, the product information can be evaluated against a set of plant processes, representing the selection of different production plants.

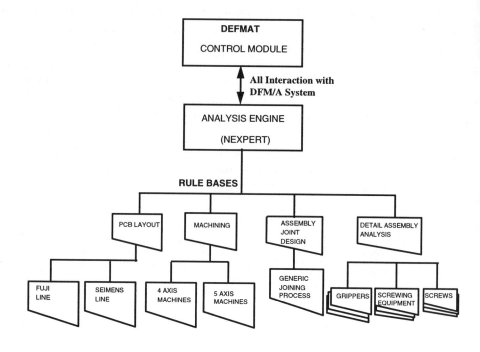

Figure 6.12 *Analysis Engine Implementation Architecture*

All rules to be used in the scenario are defined in the rules file. The rules are defined accessing objects through the class definition defined in the class file. Thus the rules can be used in any customer installation provided the customer defined the product and process characteristics according to the class definition. The flexibility of the system is demonstrated through the user selection of processes and manufacturing resources during analysis.

Generic DFM analysis

The generic DFM analysis structure is used to provide the structural framework of analysis. All analyses are addressed in the stages of :

• Analysis "make" hypothesis:
  ⇨user select or specify data elements for analysis

⇨load and verify data elements
⇨generate and verify list of all possible processes
⇨evaluate product element with process list
⇨advise user on any violations of processes and log to DEFMAT window
⇨group all acceptable processes
⇨present processes for user to select
⇨update product model database

In the machining scenario, the user selects the features from the CAD interface and they are evaluated against the designated plant process model. The system can also loop the analysis through all the features in a piece part. In the PCB scenario, the board and each components forms the data elements. Each of them are evaluated against the production line constraints.

The similar analysis structure for assemblies is used in the Detailed Joining Analysis scenario to evaluate the selection of screws and screwing equipment. The joining features are the hole and dome pairs.

- Analysis "join" hypothesis:
  ⇨user select or specify joining features on the CAD screen
  ⇨load and verify features and parent piece part
  ⇨perform geometric transformation on features location
  ⇨generate and verify list of all possible screws
  ⇨evaluate hole and dome with screw list
  ⇨advise user on any violations of screws and log to DEFMAT window
  ⇨group all acceptable screws
  ⇨present screws for user to select
  ⇨update product model database

The higher level product assembly analysis is illustrated in the Conceptual Joining scenario.

- Analysis "product joint" hypothesis
  ⇨identify significant joint requirements
  ⇨establish values for joint requirements
  ⇨form potential sets of joining processes
  ⇨for each requirement log acceptable and unacceptable processes to DEFMAT window

⇨iterate until user accepts trade-off
⇨pool all common joining processes
⇨rank processes with additional preferences
⇨present processes for user to select
⇨update product model database

There are additional analyses for non-process related design rules. These check on the material and weight characteristics of the product and log the feedback to the DEFMAT window.

6.10.3   Rules Implementation

The Nexpert™ Version 2.0 environment supports full forward and backward chaining of rules; and downwards and upwards inheritance of object properties. In the DEFMAT implementation, backward chaining is the main referencing mechanism. The high level rules of process selection and design advice is implemented through the generic problem solving paradigms of :

- sequential procedure
- algorithmic evaluation
- constraint satisfaction
- goal searching
- ranking and sorting

Sequential procedure is the main mechanism to control the sequence of analysis. The high level analysis objectives are decomposed into detail objectives. They form hypotheses that are arranged in their analysis sequence. Backward chaining of the top level hypothesis fires the rules in sequence and provides control of analysis. An example from the rule sets is:

CONDITIONS :
        Yes rank_auxiliary_part
        Yes rank_auxiliary_material
        Yes rank_manufacturing_cost
        Yes rank_material_efficiency
        Yes rank_operational_safety

    Yes   rank_recyclability
    Yes   rank_removability
    Execute "RankList"
    HYPOTHESIS : set_method_rank
    INFERENCE PRIORITY : 1

This hypothesis runs through the sequence of ranking procedures in joint selection. Looping for repeating procedure forms another part of the control mechanism in analysis.

Algorithmic evaluation is used in computing values for analysis. These computation includes numeric, string and Boolean. Nexpert™ provides a complete range of computation functions. An example of complex enumeration is the transformation of local co-ordinates of features to the absolute co-ordinates for comparison.

Constraint satisfaction is the main mechanism in setting the threshold for process selection and design advice. The comparison can be numeric, string or boolean. In process selection, the qualified processes are attached membership to the appropriate classes. The previous example in ranking methods illustrated this method. In design advice, the different cases are constructed as different rules in satisfying the same hypothesis. An example (simplified from the original Nexpert™ notation) is the checking for components in the PCB scenario:

CONDITIONS :
    <component_height>   <=   <current   PCB   pick   and   place
    device.allowable above board protrusion>
HYPOTHESIS : check_component_height
ACTIONS :
    Execute Write "Current component height is OK for the current pick
    and place equipment allowable above board protrusion."
INFERENCE PRIORITY : 1

CONDITIONS :
    <component_height> > <current PCB pick and place device.allowable
    above board protrusion>
HYPOTHESIS : check_component_height
ACTIONS :
    Execute Write "Component height is too large for the current
    equipment. Allowed range is" <current PCB pick and place
    device.allowable above board protrusion>
INFERENCE PRIORITY : 1

Goal searching is the main mechanism to realise analysis. The use of goals (hypotheses in Nexpert™) has been illustrated in the last example. Ranking and sorting is used in ordering processes for selection. An example is in the ranking methods above. Nexpert™ provides the ranking calculation methods. The diversity of the demonstration scenarios and analysis rules demonstrate the flexible nature of the DEFMAT Analysis Engine.

6.11   Conclusions

DEFMAT was started as a response to the industrial need for tools to assist design personnel in the process of concurrent engineering. It provides a generic architecture that can be used to develop tools that can be applied in a wide spectrum of design activities. The context of design activities can be characterised in three aspects: Product Domain, Design Focus and Design Manufacturing Processes.

The Product Domain defines the product type being analysed. The Design Focus defines the stage that the product is going through in its life-cycle. The stages defined can be :

- product configuration and layout (conceptual design)
- component design (embodiment design)
- detail design
- prototype
- tooling
- production
- service
- end of life

The focus provides the objectives that the particular design solution strives to address. The current categorisation focus more on the design development stages, although work in the downstream and end of life stages are gaining significance as a result of increasing environmental concerns.

The Manufacturing Processes define the context of design according to the processes and the equipment for manufacture. The DEFMAT Manufacturing Process Model extends horizontally to the different types of processes and vertically to represent process details, equipment and tools. The richness in detail corresponds to the depth of analysis. In the most general level, the analysis corresponds to good design practice. At the very detailed level, the analysis advises on the design of the component to suit the particular machine and tools in the factory floor. The implementation choice is the trade-off between the efforts in eliciting the knowledge and the benefits gained from the detailed analysis.

An industrial prototype was constructed to demonstrate a particular implementation of the generic DEFMAT architecture using current proprietary software tools. The demonstration scenario was developed to illustrate the potential application functions that could be built using the DEFMAT architecture.

6.11.1 Future Developments

Future developments of DEFMAT should become more industry-oriented on a larger scale. Specific attention needs to be given to supporting the human designer and the design process and it is very important that tool sets are provided for improving the design process and resultant products. The design regime is the area of the architecture requiring this attention. This will ensure the production of a design environment as complete as possible that places the designer firmly in control.

Many of the tools are needed to enable seamless links between the modelling of product performance and the product at various stages of its evolution. Product performance means those aspects of design that affect the product such as stress and thermal analysis. Such analysis will use classical and finite continuum mechanics (elements, differences, boundaries, etc.) and mechanism performance (linkages, drives, gears, etc.) with the ability to control the levels of commutibility between the levels of analysis modelling abstraction and the part being designed. This will require an extensive tool box - many of the parts of which are currently available - complete with interfaces and manipulation tools that will need to be developed.

One key feature to this work will be the product model and its sub-models that need to be co-ordinated but could well be distributed over a wide net. Product data management and traceability thus assume a great importance. Indications are that STEP, as it currently stands, will not be able to address the total requirements. Some current investigation suggests that object-oriented methodology can be further utilised via hypertext-type languages to enable the creation of interactive product models - a concept that further reveals the shortcomings of STEP.

The ability of the designer to work in a seamless environment and to have tools, advice and controlled direction produced by the system rather than their third parties will enable true design intent to be realised with decreased decision time.

These previous targets highlight the problems of information exchange that will be required to be addressed and communication protocols will need to be closely examined.

On the other end of the DEFMAT spectrum, advances into production and process planning are also a natural extension. These advances can also be coupled back to design and permit the specifications to be produced for new manufacturing units to handle new products.

All of these projects will require more advanced AI engines than are available at present and thus some work will need to be undertaken producing the next generation of DEFMAT tools.

Appendix A: Model of Interaction

This report starts from a zero base and develops a model of man-machine interaction. This model is then examined to determine and highlight those aspects relevant to Computer Aided Design.

Interaction - The beginnings

People express their ideas concerning design by the use of drawings, physical models and mathematical models. Rarely do they work alone, for interaction with others is vital for the refining process that is part of all design activity. The arguments will be developed with what is considered to be an example of bad interaction.

When new telephone boxes were introduced into Britain, in order to make a call, the following sequence had to be followed:

1. Pick up the receiver
2. Dial the number
3. Wait until somebody answers the call. This is signalled by a series of "pips"
4. Quickly insert the money and talk.

Originally the boxes contained the instruction "DO NOT INSERT MONEY UNTIL SOMEONE ANSWERS". This instruction caused a great deal of confusion and it was eventually changed. The reason for this confusion is that it tells the user what *not* to do. There is also the belief that sounds can be easily described by simple words. The instruction also depends upon a ***wrong assumption*** of the users beliefs.

Thus one arrives at a first rule of communication, (interaction always implies some form of communication).

- *A user is always dissatisfied with being told what not to do. He wishes to be told what to do. Therefore it is better to express negatives in a positive form.*

The issue of dialogue will be further pursued as the models of interaction are developed, for dialogue is one of the key factors of interaction.

Symmetry

The concept of symmetry is a key to good man-machine interaction. In simplistic terms this means that for every request there is a response both on the part of the user and the machine. It must be also a perceived response. If a user's request only produces an internal response on the part of the machine, this is not sufficient for the user must know that the machine is responding to the request. Similarly, if the machine requests a response and this does not occur after a period of time, the machine should also be able to query.

Symmetry does not preclude parallelism, for a user should be able to initiate another task whilst performing his current task. Likewise, because design is a group activity there will be times when the machine will require parallel inputs from users (hairs are not intended to be split on this subject - parallelism is considered here at the macro level of user time).

Thus with interaction we are also dealing with the concepts of language. The user requires a uniform syntax such that a "feeling" is developed for the interactions language. This uniformity must spread over the physical tools used for interaction. The requirement for a syntax means that ad-hoc commands each with an individual structure must be avoided. This leads to a second rule.

- *In interactive systems the relative importance of syntax is enhanced. Some applications are almost pure syntax.*

Two further important considerations should not be forgotten. Firstly in any interactive system there are (at minimum) two languages. These are the users (into the system), and the systems (out of the system). Secondly, language and syntax are not confined to words but can also encompass pictures, icons etc.. Syntax-controlled dialogues are thus a key feature of a correctly structured interactive system for they can generate prompts, menus and application control structures. Dialogue (in either direction) may be any syntax-structured collection of text, pictures or icons. A picture could be an engineering drawing.

The syntax developed, (or lack of it) has a direct bearing upon the user. This is because if a person's representation of a task is changed, you can change the way a person does the task. Of equal importance is the lexicon that is supported by the syntax. The world experience of a user may well determine how well he can use a system. This points to the wisdom (if possible) of getting the prospective user community to assist in the development of syntax and lexicon rather than depend entirely upon computer programmers.

It also means that a well designed system must be able to deal with both the naive and experienced user. This is no problem with a symmetrical system, for a part of a command (sentence) should be able to solicit a system response that allows the user to complete the command.

Representation

The previous concepts lead to the next caution. This is that different representations of the same problem may vary in difficulty. Using an example from mathematics this may be clearly illustrated as follows: if we wished to write a routine to draw a circle of radius r on a graphics device, the expression $y^2 = r^2 - x^2$ is more difficult to use that the form $x = r \sin \emptyset$, $y = a \cos \emptyset$. In a similar way the transfer between two problems (a frequent occurrence in a design environment) also depends upon the semantics of the problem representations.

With respect to CAD interactions, many systems require the user to undertake excessive mental gymnastics (for example in the drawing of arcs and tangents) when compared with actions and experience on the drawing board. Examination of the problem representation showed that the criteria used was ease of programming not ease of use or interaction.

These arguments are thus pointing to the concept of richness. If the user language is rich it allows for alternative paths to be used to reach particular solutions. To travel along alternative paths also implies control. Thus we can arrive at a further rule:

- *Interaction is a style of control.*

Interaction and control

The previous arguments then lead to the assertion that there must be, *a priori*, a general understanding of the application domain common to both user and machine. There must also be closure in the common understanding of the **current** context and goal. This closure must be maintained.

These factors and the requirement for symmetry require that the responsibility for goal achievement must be shared between user and machine.

Because the control aspects may occur in variety of domains, i.e. hardware, software, problem, user, all of which intersect the system should be adaptive. It should be adaptive to:

- the user
- the current context
- the available input/output devices.

The first requirement needs user profiles to be developed by the system as use proceeds. The second requirement is for responses to evolve as the interaction proceeds. Naturally these two requirements must have access to one another's information.

Ideally the system will then learn from the application domain but the systems must explain their assertions. All these are natural developments from adaptive control practice and theory and are well within the reach of contemporary technology. Finally, the system must allow an uncluttered transfer between the query, explanation and knowledge acquisition modes.

User/System Performance

As questions and answers have been developed from mainly the systems view, a series of questions are now developed from the users view. These questions are of value when considering system alternatives.

- **Time** How long does it take a user to accomplish a given set of tasks using the system? Time is not the sole measure of performance.
- **Errors** How many errors are made in accomplishing a given set of tasks? Are they serious?
- **Learning** How long does it take the novice user to learn how to use the system.
- **Functionality** What range of tasks can the user perform?

- **Recall** How easy is it for the user to recall how to use the system on a task that he has not performed for some time?
- **Concentration** How many things does the user have to keep in mind?
- **Fatigue** How tired do users get when using the system?
- **Acceptability** How do users subjectively evaluate the system?

Practical Considerations

The ideas so far arrived at are expressed in Figure A.1. More substance will now be given to these initial concepts commencing with the language aspects. The development of the syntax is a crucial issue and the syntax must be treated separately from the semantics and the lexicon.

A syntax-structured system implies that *all valid* sentential terms will be accepted by the system and produce a response. It is recommended that BNF notation is used to develop the syntax. Syntax diagrams, if needed for other programming purposes or in the construction of a syntax machine, may be derived from the BNF form. The syntax should be able to deal with both alphabetic and iconic forms.

Iconic forms are not necessarily pictorial replacements of single words but may embody the concepts or ideas expressed by part or whole of a sentential form expressed in alphabetic form.

The user should be able to move from one representation to the other if he so wishes: for example, if we wish to draw a circle we could be in that part of the syntax that allows us to select on icon in the form of a pair of compasses. The radius could then be set by moving a pointer on an iconic scale or by keying in the numerics or picking out the numerics from an iconic number pad. Or, rather than point to the compass icon the user could key in "draw circle" etc.. If at any point during this process the user is uncertain, he should be able to induce a response from the system that indicates his possible alternative paths to achieve command sequence closure.

In the development of the languages for interaction and reply the following are considered to be key requirements:
- The language should be easily understood by the user.
- Terms used as part of the language structure should reflect the domain in which the language is to be used.
- The basic language should be quick to learn.

- The language and its associate systems should be structured to include full options by default.

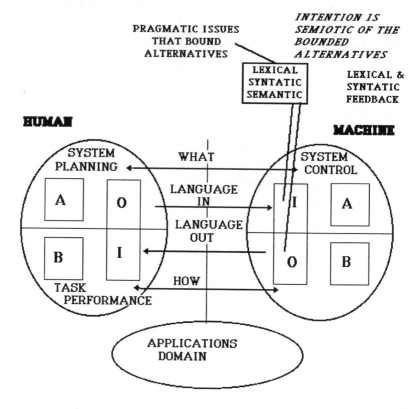

Figure A.1 *A Model of Interaction*

These requirements can be achieved by the model. The separating of the lexicon means that different lexicons can be used for different contexts (i.e. Architectural and Mechanical). A well developed syntax and iconic forms facilitate fast learning of the language. This is because the learner can appreciate the structure of a language rather than learning it blindly by rote. The defaults are only generally required at the inner level, e.g. a radius will be unity at start-up time.

Parallel input/actions can arise through obvious actions, such as the automatic dimensioning of one part of a drawing whilst creating another part of the drawing, or the not-so-obvious such as working within separate windows or the simultaneous movement of two current positions in the display (may be of value in cartographic applications).

In order for symmetry to be maintained at all times, the system should be able to re-configure itself whenever an interactive tool is added or removed. Obviously all possible tools must be known to the system.

Conclusions

This model represents one of the inputs to the development of a specification for the future CAD system. The model itself can apply to any interactive automata situation but has been illustrated by examples relating to CAD. Tools for interaction will emerge from the conjunction of the ideas in this model and the study on the nature of design and CAD that will lead to a system specification and architecture.

References

Alting, L. and Jorgensen, J. (1993) "The life-cycle concept as a basis for sustainable production", Annals of the CIRP, **42**, 1, 163-167.

Booch, G. (1994) "Object-Oriented Analysis and Design with Applications", The Benjamin/Cummings Publishing Company, Inc., USA.

Boothroyd Dewhurst Inc. (1993), "DFA/Pro", 138 Main Street, Wakefield, BI 02789, Wakefield, USA

Boothroyd, G. (1993) "Product Design For Manufacture and Assembly", The Institution of Electrical Engineers, The Second IEE International Manufacturing Lecture.

Boothroyd, G. and Alting, L. (1992) "Design for Assembly and Disassembly", Annals of the CIRP, **41**, 45-48.

Boothroyd, G. and Dewhurst, P. (1988) "Product Design for Manufacture and Assembly", Manufacturing Engineering, April, 42-46.

Bowen, J. and Bahler, D. (1991) "Supporting co-operation between multiple perspectives in a constraint-based approach to concurrent engineering", Journal of Design and Manufacturing, **1**, 2, 89-105.

Bradley, P. and Molloy, O. (1995) "A Quality Function Deployment Based Approach to Business Process Redesign", CIM At Work conference, Eindhoven, Holland, August. To appear in Computers in Industry.

Breuker, J. and Wielinga, B. (1987) "Model-Driven Knowledge Acquisition: Interpretation Models" ESPRIT Project 1098. Report, University of Amsterdam, Amsterdam, The Netherlands.

Bronsvoort, W. F. and Jansen, F. W. (1993) "Feature modelling and conversion - key concepts to concurrent engineering", Computers in Industry, 21, 61-86.

Brophy, G.S., Lettice, F.E., Sackett, P.J. and Fan, I.S. (1993) "Concurrent Engineering with CATIA", The CIM Institute, Cranfield Institute of Technology, UK.

Browne, J., Sackett, P.J. and Wortmann, J.C. (1992) "The System of Manufacturing: A prospective study", European Commission Directorate General XII , Commissioned report.

Browne, J., O'Gorman, P., Furgac, I., Felsing, W. and Deutschlander, A. (1985) "Product design for small parts assembly", Robotic Assembly, Rathmill ed., IFS Ltd., London, UK.

Business Week (1990) "A smarter way to manufacture" Business Week International, 30 April.

Canty, E. J (1987) "Simultaneous Engineering: Expanding Scope of Quality Responsibility", Digital Equipment Corporation White Paper.

Chang, T.-C. (1990) "Expert Process Planning for Manufacturing", Addison- Wesley.

Chetupuzha, J.M., Badiru, A. B. (1991) "Design considerations for knowledge acquisition", Computers and Industrial Engineering, 21, 1-4.

Clancey, W. (1985) "Heuristic Classification", Artificial Intelligence 27.

Corbett, J., Dooner, M., Meleka, J. and Pym, C. (1991) "Design For Manufacture: Strategies Principles and Techniques", Addison-Wesley Series in Manufacturing Systems, Prof. J. Browne ed., Addison-Wesley, Wokingham, UK.

Cox, B. J., Novobilski, A. J. (1991) "Object-Oriented Programming. An evolutionary approach", Addison Wesley Publishing Company, Inc., USA.

Cutting-Decelle, A.-F. and Dubois, A.-M. (1994) "Product data handled by STEP - Considerations and Proposals", Revue Internationale de CFAO. **9**, 3, 325- 337.

DARPA (1990) "DARPA Initiative in Concurrent Engineering (DICE), Mission Statements", Defence Advanced Research Projects Agency (DARPA), USA.

DEFMAT (1995) BRITE Euram Project No. 4661. Final Technical Report.

Delbressine, F. L. M. (1989) "On the integration of design and manufacturing", PhD. Thesis, Technical University of Eindhoven, The Netherlands.

Duffy, A., Andreasen, M., Bowen, J., MacCallum, K. and Reijers, L. (1994) "Design Coordination Support", Research report., Esprit Basic Research Working Group 7401: CIMDEV.

Egenton W.J. , O' Sullivan, D. (1992) "The Customer Interaction Model: A TQM Cornerstone for Software Quality Assurance", 1992, Quality Assurance Research Unit, University College, Galway, Ireland.

Engines, G. A. (1989) "DARPA Initiative in Concurrent Engineering (DICE)", Concurrent Engineering Programs, Cincinatti, Ohio, USA.

Eversheim, W. and Gross, M. (1990) "Trends and experiences in applying Simultaneous Engineering", First International Conference on Simultaneous Engineering, London, Status Meetings Ltd, UK.

Express (1990) "ISO CD 20303-11. Exchange of Product Model Data - Part 11: The EXPRESS Language".

Fearing, R. S. (1982) "Exploration of the dextrous hand control problem", G.E. Technical Information Exchange.

Feru, F., Vat, C., Timimoun, A., Coquebert, E., and Rouchon, C. (1993) "A design for manufacturing aided system based on functional and knowledge aspects", IFIP TC5 and SME joint conference on World

Class Manufacturing, September, Phoenix, USA

Gallagher, C., Lawlor-Wright T., Molloy, O. (1994) "Issues in the Development of Knowledge-Based Systems for Design For Assembly and Testability of Printed Wiring Boards", Proceedings of the Concurrent Engineering and Electronic Design Automation (CEEDA"94) Conference, Bournemouth, April, 81-86.

Glovin, S. R. and Peters, T. J. (1987) "Features as a basis for intelligent CAD/CAM", SME Technical Papers No. MS87-770.

Göbler, Th. (1992) "Modellbasierte Wissensakquisition zur rechnerunterstützten Wissensbereitstellung für den Anwendungs-bereich Entwicklung und Konstruktion", Carl Hanser Verlag, Germany.

Goldbogen, G., Hoernes, P., McCool, A. and Lim, A. (1988) "Expert Systems for Extracting Manufacturing Features from a CAD database", Expert systems and Intelligent Manufacturing, Elsevier Science Publishing, UK.

Grabowski, H., Rude, S., and Hain, K. (1994) "Core Data for Automotive Mechanical Design Processes (AP 214) - A step to STEP for the automotive industry", Revue internationale de CFAO, **9**, 3, 413-433.

Griffin, A. (1992) "Evaluating QFD's use in US firms as a process for developing products", Journal of Production Innovation Management, **9**, 171-187.

Griffin, A. and Hauser, J. (1992) "Patterns of communication among marketing, engineering and manufacturing - a comparison between two new product teams", Management Science, **38**, No. 3, March, 360-373.

Hayes-Roth, F., Waterman, D. A., Lenat, D. B. [Ed.] (1983) "Building Expert Systems" Addison Wesley Publishing Company, Inc., USA.

Henderson, M. R. (1988) "Extraction of feature information from three dimensional CAD data", PhD., Purdue University, USA.

Henderson, M. R. and Chang, G. J. (1987) "FRAPP: Automated feature recognition and process planning from solid model data", ASME

International Computers in Engineering Conference, San Diego, USA.

Hoffman, R.R. (1987) "The problem of extracting the knowledge of experts from the perspective of experimental psychology", AI Magazine, Summer Issue.

Holden, H. (1995) "DFM in PWB FAB: A Review of Predictive Engineering Benefits", Hewlett-Packard Corporation.

Honda, T., Kaneko, S. and Takeda, Y. (1993) "3-D shape reconstruction for recognition of freehand machine drawings", Annals of the CIRP, 42, 1, 185-188.

Hordvik, U. and Oehlmann, R. (1992) "Support of collaborative engineering through a shared high level product model", IFIP WG 5.7 Working Conference.

Ishii, K., Eubanks, C. and Marks, M. (1993a) "Evaluation methodology for post-manufacturing issues in life-cycle design", Concurrent Engineering: Research and Applications, 3, 1, 61-68.

Ishii, M., Tomiyama, T., and Yoshikawa, H. (1993b) "A synthetic reasoning method for conceptual design", IFIP TC5 and SME joint conference on World Class Manufacturing, September 1993, Phoenix, USA.

Jackson, S. (1991) "Qualitative modelling of unstructured knowledge to support strategy determination", 1991, Ph.D. Thesis, University College Galway, Galway, Ireland.

Jackson, S. and Browne, J. (1992) "AI-based decision support tool for strategic decision making in the factory of the future", Computer-Integrated Manufacturing Systems, 5, 2, 83-90.

Joshi, S. and Chang, T.-C. (1990) "Feature extraction and feature based design approaches in the development of design interfaces for process planning", Journal of Intelligent Manufacturing, 1, 1-15.

Jovane, F., Alting, L., Armillotta, A., Evershein, W., Feldmann, K.,

Seliger, G. and Roth, N. (1993) "A key issue in product life-cycle: disassembly", CIRP Annals: Manufacturing Technology, **42**, 2, 651-658.

Karbach, W., Linster, M. "Wissensakquisition für Expertensysteme Techniken, Modelle und Software-Werkzeuge", Carl Hanser Verlag, Germany.

Kelly-Sines, R., Lukas, M.P. and Kaya, A. (1989) "Basic concepts for integrating engineering resources to manufacture a new product", Proceedings of INCOM "89 conference.

Krause, F.L., Ulbrich, A. and Woll, R. (1993a) "Methods for quality-Driven Product Development", Annals of the CIRP, **42**, 1, 1993, 151-154.

Krause, F.-L., Kimura, F., Kjellberg, T. and Lu, S.C.-Y. (1993b) "Product Modelling", CIRP Annals 1993: Manufacturing Technology, **42**, 2, 695-706.

Kuo, W. and Hsu, J.P. (1990) "Update: Simultaneous Engineering Design in Japan", Industrial Engineering, **22**, 10, 23-26.

Liner, M. (1992) "First experiences using QFD in new product development", Design Engineering, **51**, Design For Manufacture, 57-63.

Liu, T.-H., and Fischer, G. (1993) "Developing feature-based manufacturing applications using PDES/STEP", Concurrent Engineering: Research and Applications, **3**, 1, 39-50.

Lotter, B. (1982) "Arbeitsbuch der Montagetechnik" Vereinigte Fachverlage Krausskopf-Ingenieur Digest, Mainz, Germany.

McKay, A., Bloor, M.S., and Owen, J. (1994) "Application Protocols: a Position Paper", Revue internationale de CFAO, **9**, 3, 377-389.

McMahon, C., and Browne, J. (1993) "CADCAM - From principles to practice", Addison-Wesley, Wokingham, UK.

Manoharan, C., Wang, H.-P., and Soom, A. (1990) "An expert system for design for robotic assembly", Journal of Intelligent Manufacturing, 1, 17-29.

Mantyla, M. (1990) "A modeling system for top-down design of assembled products", IBM Journal of Research and Development, 4, 5 September, 636-659.

Mantyla, M., Lagus, K., and Laako, K.. (1994) "Application of constraint propagation in part family modelling", Annals of the CIRP, 43, 1, 129-132.

Meier, A. (1991) "Advantages of using features to integrate product and process modelling - results of IMPPACT (Esprit 2165)", Esprit CIME Conference.

Meylemans, P., De Wachter, L., Detollenaere M. (1990) "New CAD CAM solution for complex pinion shafts" ASME Conference on Power Gearing, Chicago

Miyakawa, S. and Ohashi, T. (1986) "The Hitachi Assemblability Evaluation Method", First International Conference in Product Design For Assembly, 1-13.

Molloy, O. (1992) "Towards an Architecture for Design for Manufacture", Design and Computer Integrated Manufacturing Symposium, Brussels, Commission of the European Communities, Brite-Euram Programme, 10-11.

Molloy, O (1995) "A Design Environment for Concurrent Engineering", 1991, Ph.D. Thesis, University College Galway, Galway, Ireland.

Molloy, O. and Browne, J. (1993) "A knowledge-based approach to design for manufacture using features", "Concurrent Engineering: Contemporary Issues and Modern Design Tools", Parsaei and Sullivan ed., Chapman and Hall, London, UK.

Moriwaki, T. and Nunobiki, M. (1994) "Object-oriented design support system for machine tools", Journal of Intelligent Manufacturing, 5, 1, 47-54.

O'Grady, P., Kim, C., Young, R. (1992) "Issues in the Testability of Printed Wiring Boards", Journal of Electronics Manufacturing, No. 2, 45-50.

O'Grady, P., Ramers, D., and Bowen, J. (1988) "AI Constraint Nets Applied to Design for Economic Manufacture and Assembly", Computer-Integrated Manufacturing Systems, **1**, 4, 204-209.

O'Connor, L., Partridge, D., Seely, B., Guthmiller, W., and Lovette, K. (1992) "The SeeQFD Software: An environment for QFD", Worldwide passenger car conference and exposition, Dearborn, MI, USA, 1992, BAE, Warrendale, PA, USA, 17-37.

Oh, J. S., O'Grady, P., and Young, R.E. (1991) "An artificial intelligence constraint network approach to design for assembly", Technical Report, Dept. of Industrial Engineering, NCSU, USA,

Olesen, J. (1992) "Concurrent Development in Manufacturing - based on dispositional mechanisms", PhD., Institute for Engineering Design, Technical University of Denmark.

Pahl, G., and Beitz, W. (1986) "Konstruktionslehre", Handbuch fur Studium und Praxis. Berlin, Heidelberg, New York:Springer-Verlag, USA.

Parmley, R. (1977) "Standard Handbook of Fastening and Joining", McGraw-Hill Book Company, New York, USA.

Perry, S. (1992) "Real-life, reusable QFD", Electro/92, Hynes Convention Center, Boston, MA, May 12-14. Region 1 Central New England Council, METSAC, IEEE, New England and New York Chapters, ERA. 423-426.

Philippi, T. (1991) IIT Research Institute Frontiers Dec.,**15**, 12.

Pugh, S. (1991) "Total Design", Addison-Wesley, Wokingham, UK.

Reddy, Y.V., Wood, R.T., and Cleetus, K.J. (1992) "The DARPA Initiative in Concurrent Engineering", Concurrent Engineering

Research in Review, 1, 2-10.

Rumbaugh, Blaha, Premerlani, Eddy and Lorensen (1991) "Object-Oriented Modelling and Design", Prentice-Hall International Editions, USA.

Salisbury, J.K., and Craig, J.J. (1982) "Articulated hands: force control and kinematic issues", International Journal of Robotics Research, 1, Winter, 4-17.

Scholz-Reiter, B. (1992) "CIM Interfaces - Concepts, standards and problems of interfaces in Computer Integrated Manufacturing", Chapman and Hall, London.

Seliger, G., Kruger, S., and Wang, Y. (1992) "Integrated Information Modelling for Simultaneous Assembly Planning", Institute for Machine Tools and Manufacturing Technology, Chair of Assembly Technology, Pascalstrasse 8-9, D- 1000 Berlin 10, Germany.

Singh, K. J. (1992) "Concurrent Engineering pilot project at GE Aircraft Engines", Concurrent Engineering Research in Review, 4, Autumn, 14-23.

Sivard, G., Lindberg, L., and Agerman, E. (1993) "Customer-Based Design with Constraint Reasoning", Annals of the CIRP, **42**, 1, 1993, 139-142.

Sobolewski, M. (1990) "Dicetalk - An object-oriented knowledge based engineering environment", 5th International Conference on CAD/CAM, Robotics and Factories of the Future. Norfolk, Dec. 2-5, UK.

Spath, D. (1994) "The utilisation of hypermedia-based information systems for developing recyclable products and for disassembly planning", Annals of the CIRP, **43**, 1, 153-156.

Spooner, D. L. (1994) "An object-oriented product database using ROSE", Journal of Intelligent Manufacturing, 5, 1, 13-21.

Stoll, H. W. (1988) "Design For Manufacture", Manufacturing

Engineering, January, 66-73.

Sturges, R.H. and Wright, P.K. (1989) "A quantification of dexterity: the Design for Assembly calculator", Robotics and Computer Integrated Manufacturing, **6**, 3-14.

Terry, W.R., Karni, R., and Richards, C.W. (1990) "A knowledge based system for the integrated design and manufacture of round broach tools", Journal of Intelligent Manufacturing, **1**, 77-91.

Thackeray, R., and Treeck, G.V. (1990) "Applying Quality Function Deployment for Software Product Development", Journal of Engineering Design, 1, 4, 1990, 389- 410.

Tipnis, V. A. (1993) "Evolving issues in product life-cycle design", Annals of the CIRP, **42**, 1, 169-173.

Tipnis, V.A. (1994) "Challenges in product strategy, product planning and technology development for product life-cycle design", Annals of the CIRP, **43**, 1, 157- 162.

Tomiyama, D.T. (1992) "The Technical Concept of Intelligent Manufacturing Systems (IMS)", The University of Tokyo, Tokyo, Japan.

Tomiyama, T., Umeda, Y., and Yoshikawa, H. (1993) "A CAD for Functional Design", Annals of the CIRP, **42**, 1, 1993, 143-146.

Venkatachalam, A.R., Mellichamp, J.M., and Miller, D.M. (1993) "A knowledge-based approach to design for manufacturability", Journal of Intelligent Manufacturing, 4, 355-366.

Vincoli, J.W. (1993) "Basic Guide to System Safety", VNR Basic Guide Series, USA.

Wallace, D.R., and Suh, N.P. (1993) "Information-Based Design for Environmental Problem Solving", Annals of the CIRP, **42**, 1, 1993, 175-180.

Weule, H. (1993) "Life-Cycle Analysis - A strategic element for future products and manufacturing technologies", Annals of the CIRP, **42**, 1, 181-184.

Yoshikawa, H. (1981) "General Design Theory and a CAD System in Man Machine Communication in CAD/CAM", North Holland, The Netherlands.

Yoshikawa, H. (1993) "Systematisation of Design Knowledge", Annals of the CIRP, 42, 1, 131-134.

Yoshikawa, H., and Warman, E.A. (1987) "Design Theory for CAD", North Holland, The Netherlands.

Zhang, H.-C., and Alting, L. (1993) "Computerised manufacturing process planning systems", Chapman and Hall, London, UK.

Glossary

Analysis Engine   see Inference Engine.

Artwork   The lines and shapes that form design elements, such as pads and conductors, and are produced as etch patterns on the printed wiring board during manufacturing.

Assemble   The totality of operations to put together a number of geometrically defined parts. Shapeless assistant material may be used. Assembly operations can be divided into joining, handling, testing and special operations.

Assembly   A group of components of a product. The grouping is created to satisfy a functional or packaging need.

Assembly Feature   A sub-type of "feature" - elements of piece parts affecting the assembly process by determining constraints. Examples: a gripping face, a screw thread or a clip.

Assembly Machine   A machine designed to perform a particular assembly operation, e.g. a soldering machine or a riveter.

Assembly Process   The procedure of assembling a product or sub-assembly. An assembly process is determined by product, joining method, used equipment, and organisational aspects like operation sequence and structure.

Assistant Material   A material used to improve the quality of an assembly operation, e.g. coolant, or to realise a joining connection between piece parts of a product, e.g. adhesive or solder paste.

Assistant Part   A part necessary to realise a joining connection between piece
                         parts of a product, e.g. a screw or a rivet.

Attribute   A property, quality or feature that belongs to components or
                         products. For example, material type.

Bill of Material   A generic description of a product in terms of a listing of its
(BOM)   components.

Board   In electronics manufacturing, board usually refers to the
                         "production unit," or panel, that goes through module
                         assembly. A single board in module assembly may hold one or
                         more circuits, depending on the manufacturing site and circuit
                         size.

Circuit   A term that refers to a network or a portion of a network. The
                         term is also commonly used to refer to all of the conductors
                         and nodes on the design. In manufacturing, circuit refers to a
                         functional printed wiring-board that has undergone
                         manufacturing processes.

Class   An abstract description of a collection of objects which share
                         at least one common distinctive mark.

Component   A constituent part or aspect of a more complex product.

Connecting   Equipment used to realise a mechanical, kinetic or
Equipment   informational linkage between at least two components of a
                         system.

Construction   A sub-type of "feature". Construction features are function-
Feature   oriented features, e.g. a drilling hole or a feather groove.

Data Model   Formal method to present information in a structured way from
                         a defined, delimited field of information.

Design for Assembly The design of components and products paying particular
                         attention to reducing assembly time and costs.

Discrete
A device used as a distinct and individual component and packaged as a stand alone element. Diodes and small outline transistors (SOTs) are examples.

Element (Technological)
Part of a product; typically it is parameterised and defined as an element with both geometrical, dimensional and manufacturing information linked to it.

Expert System
Specific program describing knowledge and running a reasoning engine. It can either be bought as a standard product or specifically developed for a certain application, using either special AI languages (e.g. LISP, PROLOG) or standard programming languages (e.g. C(++), Pascal).

Expert System Shell
Programming environment in which to develop expert systems.

Feature
A geometric element of a piece part logically connected with semantic meaning. Geometric elements are for example surfaces, solids or groups of contours. The semantic meaning is expressed by static or dynamic attributes. A dynamic attribute can be a program. Features can be classified into assembly, construction, and manufacturing features.

Fiducial
A visual reference target created on a design for surface mount devices and products. Manufacturing use fiducials for board and component alignment.

Fit
A fit is created by the relation of the tolerance fields of two joined parts.

Footprint
The set of pads that allow the pins of a surface mount component to make electrical contact with the board.

Function
General connection between input and output quantities of a system. It embodies the objective to perform a specific task.

Gripping Face
A face of a component which is used for gripping the component.

Handling features   Features which allow easy handling of a component. For example, when using robotics it is essential that the part has flat gripping faces.

Inference Engine   Component of an expert system which performs the general problem solving.

Instance   A representative of a class.

Interface (CAD)   Means of accessing a CAD system and its data, either interactively using mail box or pipes, or using an intermediate storage medium e.g. database or flat file.

Joining   A sub-type of "manufacturing method". Joining is defined as putting together two or more workpieces of geometrically defined shape or of workpieces with shapeless material.

Joining Face   A face of a piece part that is in touch with a joining face of another piece part (joining partner) during or at the end of an assembly process.

Joining Material   Material used to sustain joined components.

Joining Partner   Two or more piece parts connected by a joining relation.

Joining Relation   Junction between at least two piece parts realised by joining.

Joint   A junction of two or more components. Also, a kinematic pair of two rigid members (links) of a kinematic structure enabling one to have a motion in relation to the other.

Kinematic Structure   Topological structure of objects represented by joints and links.

Knowledge Acquisition   Gaining knowledge, usually from an expert.

Knowledge Base   Collection of domain knowledge. It contains facts represented according to a defined formalism.

Layout Two- or three-dimensional outline for visual description of the arrangement of system components regarding building and transport constraints.

Manufacturing Feature A sub-type of "feature" which describes elements of piece parts, which are ensured to be manufacturable when they are defined. For example a pattern of drilling holes. Manufacturing features can be used for operations scheduling and NC-programming.

Onsertion The process of placing surface mount components on the board.

Operation Action that can be decomposed into a set of operation elements resulting in a manufacturing progress.

Operation elements Part of an operation which usefully cannot be further decomposed.

Operator A competent person designated to start, monitor and stop the intended productive operation of a resource.

Orientation Positioning in relation to a specific direction.

P.C.B. Printed Circuit Board.

Piece part Individual component of a product that is not decomposed into constituent parts.

Pitch The distance between two or more corresponding points such as the distance between pins on a component, or the distance between the centres of adjacent conductors.

Process characteristics By process characteristics we mean characteristics which influence what products and components can be assembled using the process.

Process constraints Factors of the process which inhibit design of products or components.

Product Domain   Products having similar specific characteristics.

Product Structure   The arrangement and interrelationships of components in a product.

Surface Mount   A technology that places parts on a board by attaching
Technology (SMT)   component leads to the module without the use of intrusive pins.

Test Point   A point such as a pin, via, or test pad where a test probe may be placed to inject or sample a signal. Test points are probed during the manufacturing process to detect short circuits, broken connections, and other electrical faults on a circuit.

Through-Hole   A hole that passes from one board side to another. The term through via is frequently, but inappropriately, used as a synonym.

Tool   An instrument used for automated performing of operations, e.g. gripper, welding tong or soldering iron.

Index